Practical Soldiers

History of Warfare

VOLUME 107

The titles published in this series are listed at *brill.com/hw*

Practical Soldiers

Israel's Military Thought and Its Formative Factors

By

Avi Kober

BRILL

LEIDEN | BOSTON

Cover illustration: The tank overrunning a book is an illustration of the IDF's attitude towards military theory. © Shahar Kober.

Kober, Avi, author.
Practical soldiers : Israel's military thought and its formative factors / by Avi Kober.
pages cm
Includes bibliographical references and index.
Summary: "This book suggests a general framework for the analysis of formative factors in military thought and offers an account of the Israel Defense Force's state of intellectualism and modernity. This account is followed by an attempt to trace the factors that have shaped Israeli military thought. The explanations are a mixture of realist and non-realist factors, which can be found at both the systemic and the state level of analysis. At the systemic level, realist evaluations focus on factors such as the dominance of the technological dimension and the pervasiveness of asymmetrical, low-intensity conflict; whereas at the state level one can find realist explanations, cultural factors, and societal influences. Moral and legal constraints also factor into both the systemic and state levels", Provided by publisher.
ISBN 978-90-04-30653-0 (hardback : alk. paper) -- ISBN 978-90-04-30686-8 (e-book) 1. Israel--Armed Forces--History--21st century. I. Title.

UA853.I8K63 2015
355'.03355694--dc23

2015035252

This publication has been typeset in the multilingual "Brill" typeface. With over 5,100 characters covering Latin, IPA, Greek, and Cyrillic, this typeface is especially suitable for use in the humanities.
For more information, please see www.brill.com/brill-typeface.

ISSN 1385-7827
ISBN 978-90-04-30653-0 (hardback)
ISBN 978-90-04-30686-8 (e-book)

This book is printed on acid-free paper.

Contents

Preface IX
The Research Question and Main Arguments XI
Methodology XIII
Sources XVI
Structure XVI
Acknowledgements XVII
List of Tables XVIII

1 **Military Thought and the Formative Factors Affecting It** 1
The Intellectual and Modern Focus of Military Thought 1
Intellectual Focus 1
Modern Focus 4
Realist vs. Cultural Formative Factors 5
Is There Such Thing as a Universal Strategic Culture orMilitary Culture? 7
Formative Factors: The Systemic Level 9
Technology as a Dominant Dimension 9
The Broadening of War 11
The Narrowing of War 17
The Complexity of War 22
The Pervasiveness and Importance of LICs 23
Zeitgeist 25
War in a Given Period 25
Intellectual Climate 28
Morality, Law, and War 30
The Unit (State) Level 31
The State's Specific Strategic Conditions 31
Organizational Factors 32
Lesson Learning Processes 34
The Level of the Individual 35
Thinkers' Operational Code: Fatalists, "Military Intellectuals," "Intellectual Soldiers," And "Practical Soldiers" 36
Spiritual Fathers 38
Personal Experience 39
Conclusion 40

2 **Israeli Intellectual and Modern Focus** 43
Intellectual Weakness 43
Symptoms of Poor Intellectualism in the Military 43
A Number of Positive Symptoms 57
The Detrimental Effect of the IDF's Lack of Intellectualism 69
The Cost of Poor Intellectualism 69
The Danger Entailed in Emulated Doctrines 72
Damage Inflicted by Phony Intellectualism 73
Modern Focus 75
Types of Conflict 75
Levels-of-War 76
Dimensions of Strategy 79
Post-heroic Mindset and Moral and Legal Considerations Becoming an Integral Part of Israeli Military Thinking 82
Conclusion 84

3 **Systemic Formative Factors** 86
The Pervasiveness of LICs 86
Technological Developments 92
The Ascendancy of Firepower 92
The Notion of "Small But Smart Military" 99
Victory from the Air 100
Network-centric Warfare 103
Command and Control over Plasma Screens 104
Cyberspace and Cyber Warfare 105
Legal and Moral Constraints Created by International Law and Norms 107
Jus ad Bellum: Preventive War as Opposed to Preemptive Attack 107
Jus in Bello 109
Conclusion 111

4 **Unit (State)-Level Factors** 114
Geostrategic Factors and the Cult of the Offensive 114
Cultural Factors 115
"Bitzuism" (*A Performance-oriented Approach*) *and Experience-based Intuition* 115
Extolling Resourcefulness and Improvisation 118
The Post-1967 Hubris and Intellectual Feebleness 122
Slow Adaptation to a LIC Mindset 123
Lack of Institutional Intellectualism 126

Politization and Militarization Processes 127
Societal Factors 128
An Occupational – Rather than Institutional – Military 128
New Elites and Their Impact of the IDF's Military Thought 129
The Near-lost Competition with the Civilian Sector 130
Explanations for Israeli Post-heroic Mindset's First Rule 132
Commitment to Jewish and Western Democratic Moral and Legal Standards 132
"Purity of Arms" 133
Justifying the Use of Force 134
The "Judicialization" of Israeli Military Thinking 136
Conclusion 137

5 **Conclusion** 138
General Formative Factors 138
The Israeli Case 142
The State of Israeli Military Thought: Intellectual and Modern Focus 142
Formative Factors 143
What Can Be Done to Improve the Level of Intellectualism of Israeli Military Thinking? 145

Appendix 149
Bibliography 162
Index 184

Preface

Decades of Israeli independence and Arab-Israeli wars have not produced a single comprehensive study which is entirely dedicated to Israeli military thought. What the literature has to offer is a number of partial accounts of Israeli military thought that cover specific aspects thereof. Just to mention a few examples: In their analysis of the "the revolution in Israeli national security affairs" Eliot Cohen et al. wrote about the IDF's anti-intellectual state of mind and the impact of the American Revolution in military Affairs (RMA) on Israel's strategic doctrine.[1] Ariel Levite analyzed offense and defense in Israeli military doctrine.[2] A more recent work by Dima Adamsky discussed the differences in the adoption and implementation of the RMA between the Soviets, the Americans and the Israelis as a result of different cultures, portraying the Israelis as intellectually weak but good at improvisation and practice.[3] Eitan Shamir showed how the attitudes of the IDF to the doctrine of mission command were affected by Israel's strategic culture and other factors such as technological developments and the pervasiveness of LICs.[4] Another work by Shamir discussed the Americanization of the IDF.[5] Previous studies by the author of this book have also portrayed a partial picture of Israeli military thought. One work analyzed articles published in *Maarachot*, the IDF's professional journal, from 1948 to 2000, based on the assumption that these articles were a mirror of the IDF's officer corps' intellectual qualities and professional areas of interest.[6] Other works focused on the changes Israeli military thought has gone through over the years;[7] described and explained the rise and fall of

1 Eliot A. Cohen, Michael J. Eisenstadt, and Andrew J. Bacevich, *Knives, Tanks, and Missiles: Israel's Security Revolution* (Washington, DC: Washington Institute for Near East Policy, 1998).

2 Ariel Levite, *Offense and Defense in Israeli Military Doctrine* (Tel Aviv: Jaffee Center for Strategic Studies, 1988).

3 Dima Adamsky, *The Culture of Military Innovation* (Stanford: Stanford University Press, 2010).

4 Eitan Shamir, *Transforming Command: The Pursuit of Mission Command in the US, British, and Israeli Armies* (Stanford: Stanford University Press, 2011).

5 Eitan Shamir, "When Did a Big Mac Become Better than a Falafel? Israel's New Way of War, Military Emulation Theory and the Americanization of the IDF," an unpublished paper.

6 Avi Kober, "Israeli Military Thinking as Reflected in *Maarachot* Articles, 1948–2000," *Armed Forces & Society*, Vol. 30, No. 1 (Fall 2003), pp. 141–160.

7 Avi Kober, "What Happened to Israeli Military Thought?" *Journal of Strategic Studies*, Vol. 34, No. 5 (October 2011), pp. 707–32.

Israeli operational art;[8] offered an account of Israeli attrition strategy;[9] and discussed the role played by battlefield decision in Israeli security conception and practice.[10] Scattered and sporadic references to Israeli military thought can be found in studies dedicated to the history of the IDF, e.g., books by Luttwak and Horowitz[11] or Martin Van Creveld.[12] This book offers a broader perspective of the characteristics of Israeli military thought and its formative factors, which is based on a general framework for the analysis of the formative factors in modern and post-modern military thought, which is provided in Chapter 1.

The title of the book, *Practical Soldiers*, comes from a term coined by Basil H. Liddell Hart.[13] It has a very negative connotation, referring to commanders who put their faith in experience and intuition rather than in any intellectually-acquired knowledge. Practical soldiers believe that in war, like in love, one cannot understand what it is all about until one undergoes the experience in person.

Throughout the years, the IDF has shown symptoms of poor intellectualism, which has been reflected in a lack of interest in the theoretical aspects of the military profession and an underestimation of the contribution of theory to practice. This has had a detrimental effect on the IDF's performance. Since the 1990s, the IDF has been emulating the RMA-inspired American doctrine, which has come at the expense of its originality and innovation, and as if to add insult to injury, Israeli military thinking has been affected by false intellectualism and intellectual pretense.

The book identifies additional characteristics of Israeli military thought. First, a late accommodation to LIC challenges, and a negative effect of policing missions in the occupied territories on the IDF's military thought and performance in LIC situations. Second, a strong tactical orientation, which, in recent years, has gradually been mitigated by a greater interest in the operational and grand-strategic levels as well as an understanding of the two-way relationship between the two extremes of the levels-of-war pyramid (the so-called tacticization of strategy as opposed to the strategization of tactics). Third, a cult of

---

8 Avi Kober, "The Rise and Fall of Israeli Operational Art," in Martin van Creveld and John A. Olsen (eds), *Operational Art: From Napoleon to the Present* (New York: Oxford University Press, 2011), pp. 166–94.

9 Avi Kober, *Israel's Wars of Attrition* (New York: Routledge, 2009).

10 Avi Kober, *Battlefield Decision in the Arab-Israeli Wars, 1948–1982* (Tel-Aviv: Maarachot, 1995) [Hebrew].

11 Edward Luttwak and Dan Horowitz, *The Israeli Army* (London: Allen Lane, 1975).

12 Martin Van Creveld, *The Sword and the Olive* (New York: PublicAffairs, 1998).

13 Basil H. Liddell Hart, *Thoughts on War* (London: Faber, 1943), pp. 96–7.

the offensive, which has been replaced by a more balanced approach to offense and defense as result of the ascendancy of firepower over maneuver and the requirements of conducting LICs. Fourth, a strong technological orientation, which has gradually taken the form of a cult of technology. Another, relatively new feature of Israeli military thought has been the centrality of moral and legal considerations, which have become an integral part of Israeli military thought.

A comparison of *Maarachot* and the American journal *Military Review* articles shows that as far as the modern focus is concerned the two journals reflect similar changes in war and strategy. When it comes to the intellectual focus, however, the impression one gets from the comparison is that Israeli military thought lags behind American military thought.

The Research Question and Main Arguments

Given these characteristics of Israeli military thought, the main question this book seeks to address is: What have been the factors that have accounted for the state of Israeli military thinking? The answer the book offers to this question is to be found at the systemic and the unit (state) levels-of-analysis.

The main systemic factors have been the ascendancy and pervasiveness of LICs; technological developments; and international, particularly Western moral and legal norms. Since the mid-1980s, LICs have become Israel's basic strategic threat, and have required adapting to the new reality at the levels of both military thought and practice. The change has manifested itself in the following aspects: A gradual understanding that the era of *Blitzkrieg* was over and that Israel will henceforth experience blow-for-blow confrontations that may last years; awareness – as a result of the Intifadas – of the nature and meaning of asymmetrical conflicts and attrition situations; a more balanced approach to the role played by offense and defense, which has mitigated the ultimate commitment to offense in Israeli HICs; a belief that decisive victory was no longer feasible (which was refuted in the First Lebanon War); emulation of imported concepts from the US, such as the notion of "leverages and effects;" and the adoption of post-modern ideas, such as a "victory image" as a substitute for the material/physical aspects of military decision. After the IDF's poor performance during the 2006 Second Lebanon War, which was attributed, at least partially, to its American inspired operational conception, the IDF became more critical of American doctrine.

Another influential systemic formative factor has been technological developments. In the past, Israeli senior commanders were aware of the danger entailed in over-reliance on technology at the expense of the human factor. In recent decades, however, strongly inspired by technological developments, technology has started overshadowing the non-material aspects of Israeli strategy and tactics, becoming the main factor in Israeli military thought, buildup and operations. A cult of technology has gradually developed, predicated on the belief that thanks to the unprecedented availability of precise, long-range, highly-destructive weapons, information dominance, and new means of command and control, it has now become possible to dramatically reduce the fog of war, to reduce casualties and collateral damage, and to kill without confronting the enemy face-to-face. For Israel – a state suffering from a strong sense of quantitative inferiority, which sought any force multiplier for compensation – a technology-based military was found to be a very appealing solution.

A third, entirely different systemic factor, which has gained tremendous relevance for, and impact on, Israeli military thought, has been international, particularly Western, moral and legal norms. Ethical dilemmas in war have intensified due to two main developments: the prevalence of asymmetrical conflicts and the unprecedented destructive and murderous nature of terror. Israel, like any other Western democracy, is obliged to abide by international law and liberal norms. Yet Israel has had to fight, often in complex, morally-challenging situations, contrived by terrorists devoid of moral inhibitions. Almost without explicitly acknowledging it, issues of just war, discriminate use of force, proportionality and civil liberties have permeated Israeli military thought, becoming an integral part thereof, and strengthening the intellectual aspects of military thought from an unexpected direction.

Factors at the unit (state) level have shaped Israeli military thought as well. The most realist factor at this level has been Israel's basic strategic conditions, particularly its geo-strategic ones, i.e., its tiny size and lack of strategic depth, and its quantitative inferiority vis-à-vis an Arab war coalition. Like pre-World War II Prussia and Germany, which had been operating under similar geo-strategic circumstances, Israel's conditions have dictated a force multipliers-oriented military thought, which emphasized offense, indirect approach, concentration of forces and fire, first strike, and *Blitzkrieg*, in the hope of compensating for these conditions. But the deeper the commitment to force multipliers, the higher the chances they might turn into cults, such as the cult of the offensive, or the cult of technology. In turn, reality and its challenges have had a sobering effect on these cults, and, at least in the case of the former, have resulted in a more balanced approach to offense and defense.

Cultural factors at the unit level have been detrimental to the development of Israeli military thought, and have been reflected in its a- or anti-intellectual approach: the Sabras' (native-born Israelis) performance-oriented approach and their tendency to rely on experience-based intuition; the tendency to extol resourcefulness and improvisation; the post-1967 hubris; slow adaptation to a LIC mindset; lack of institutional intellectualism; and politization and militarization processes that have left military thought in the hands of the military, without making sure that it enjoys sufficient intellectual basis.

A societal explanation stresses the following factors: the occupational rather than institutional military; the emergence of new military elites; the near lost competition with the civilian factor; and a greater casualty aversion in conflicts that do not involve the most vital interests of the state, let alone its survival.

Finally, the upholding of moral and legal considerations, stemming, at the unit level, from the Israeli society's Jewish and liberal norms, has had a considerable impact at the unit-level. Tension has developed between law and ethics, on the one hand, and the need to ensure operational effectiveness, on the other, and effort has been made in recent decades to bridge the two, by adopting a post-heroic way of war.

The absence of institutional intellectualism in the Israeli military establishment notwithstanding, islands of military thought have existed thanks to "intellectual soldiers," who, while in service, used to study, reflect and write on military matters; and "military intellectuals," whose nexus with military thought has been affected either by their affiliation with the military as staff officers or as non-uniformed experts, who worked with, for, or alongside the military, or outside the military, as academics. In a few cases, contributions of military intellectuals have been detrimental to military thought, as in the case of the IDF's Operational Theory Research Institute (OTRI), whose heads embraced non-military post-modern theories while neglecting the classical foundations of military thought.

Methodology

In order to profile Israeli military thinking and to evaluate the IDF's level of intellectualism and modern focus, theory-based criteria are employed. These criteria are accompanied by a content analysis of *Maarachot* publications – both articles and books (*Maarachot* is the name of both the IDF's professional journal and the IDF's publishing house). The content analysis covers a total of 3,585 articles and 370 books that were published between 1948 and 2008.[14]

14 On qualitative content analysis, see for example, Robert P. Weber, *Basic Content Analysis* (Newbury Park: Sage Publications, 1990); Klaus Krippendorff, *Content Analysis: An introduction to Its Methodology* (Beverly Hills: Sage Publications, 1980). For application to

Maarachot publications have been chosen for three main reasons: (a) Books and articles published by *Maarachot* could be considered a reflection of Israeli military intellectual qualities and professional areas of interest. (b) They are accessible to the public. (c) Unlike any other Israeli source, *Maarachot* is the only entity that has regularly and uninterruptedly covered Israeli military thinking from the pre-State period (*Maarachot* was established in 1939) to the present. As such, *Maarachot* publications enable us to identify trends, emphases, and changes in Israeli military thought throughout the years.

The deductive, qualitative content analysis technique applied here for portraying Israeli military thinking compresses large volumes of data into *a-priori* formulated variables. The comparison of frequencies of the selected categories enables us to make specific inferences from the data regarding Israeli military thought and shifts and changes thereof over time.

The basic criterion in the *intellectual focus* relates to the degree of history-based theoretical discussion, as compared to the more practical fields of thought about security and war, i.e., doctrine and planning. Three variables are used for measuring *modern focus*: types of conflict referred to, levels-of-war related to, and dimensions of strategy addressed by authors. Types of conflict range from unconventional, via conventional to sub-conventional. The levels-of-war range from tactics (engagements and battles), via the operational level (campaigns, operations) and strategy (the upper military level), to grand-strategy (war as a multi-dimensional activity, or war conducted at the national level). The dimensions of strategy are divided into operational, technological, logistical and societal.

As far as authorship is concerned, authors' status is divided into three categories: regulars (those in active service, most of them career officers); reservists and retired; and civilians. *Rank* is divided into two categories – NCOs and officers – with the officer sub-category divided again into ranks. Authors' country affiliation is divided into Israelis, non-Israelis, and unknown.

Systematic studies on journals' authorship or contents have been conducted for social science journals, either for single journals,[15] or for comparative purposes. Most of these have focused on authorship rather than content, and have

article authorship and content, see for example DR Dillon et al., "Article Content and Authorship Trends in *The Reading Teacher*, 1948–1991," *The Reading Teacher*, Vol. 45, No. 5 (1992), pp. 362–8.

15 For article analyses with a focus on security matters, see: Ralf Zoll, "The Journal *Armed Forces & Society*: Contributions and Contributors," paper presented at the Biennial Meetings of the Inter-University Seminar on Armed Forces and Society, Baltimore, Maryland, 20–22 October, 1989); Morten G. Ender, "Authorship and Affiliation in *Armed Forces & Society*: Volumes 1–25," *Armed Forces & Society*, Vol. 27, No, 4 (Summer 2001), pp. 623–38;

been particularly interested in single vs. multiple authorship, gender difference, and institutional or discipline affiliation.[16] In *Maarachot*'s case, analyzing similar aspects would be either irrelevant or unnecessary as, unlike purely academic publications, the question whether *Maarachot* authorship is characterized by collaboration or not has very little impact on the quality or content of the military thought expounded by the authors, if at all. The *Maarachot* survey also puts the greatest emphasis on the contents of articles and books rather than authorship, although the latter is addressed, too.

The survey is divided into sub-periods: from the 1948 Israeli War of Independence to the 1956 Sinai War; from 1957 to the 1967 Six-Day War; from 1968 to the 1973 October War; from 1974 to the 1977 Sadat initiative; from 1978 to the 1982 First Lebanon War; from 1983 to the outbreak of the First Intifada in 1987; from 1988 to 1994 (the 1993 Oslo Accords and the peace with Jordan in 1994); from 1995 to the outbreak of the Second Intifada in 2000; from 2001 to 2004 (the Second Intifada years); and from 2005 to 2008.

In order to find out whether there was a correlation between the trends identified in *Maarachot* publications and those in journals abroad, or if these trends have been unique to Israel, a comparison is made with the American journal *Military Review*. This journal shares similar orientation with *Maarachot*: Like *Maarachot*, it is committed first and foremost to covering land warfare issues, but it is also interested in the wider picture, and in a variety of topics related to changes in the nature of war and its conduct. The *Military Review* survey employs the same categories that are employed in the *Maarachot* survey, but it is partial and limited to two selected periods – 1982–1987 and 1995–2008 – which cover the Cold War's last years and its aftermath.

The formative factors in Israeli military thought are divided into two levels-of-analysis, the systemic level and the unit (state)-level. These levels include both realist and non-realist, cultural explanations. No special chapter is devoted to the level of the individual, although Israeli individuals have been active in doctrinal matters over the years, thereby balancing, at least to some extent, the image of Israeli commanders as practical soldiers. But in the absence of outstanding Israeli masters of war or spiritual fathers, I preferred referring to individuals in the framework of the discussion of the role played by "intellectual soldiers" or "military intellectuals."

Nils P. Gleditsch, "Focus On: *Journal of Peace Research*," *Journal of Peace Research*, Vol. 26, No. 1 (1989), pp. 1–5.

16 For a detailed list of such publications, see Ender, "Authorship and Affiliation in Armed Forces & Society: Volumes 1–25," pp. 636–7.

Sources

The sources of the book are both primary (e.g., documents, declarations, surveys, interviews) and secondary (researches, books and articles analyzing various aspects of military thought, both general and Israeli). Although data on the past is gradually becoming declassified and research relating to Israel's early periods has been conducted and published in recent years, it is still much easier to collect data on the more recent cases, thanks to the multitude of available and accessible information sources in recent decades – books, articles, internet sites, and media sources.

Structure

The book starts with a presentation of the general formative factors in modern and post-modern military thought, which serve as a framework for the analysis of the Israeli case. It then describes the intellectual and modern focus in Israeli military thought, and offers explanations at the systemic and the unit (state) levels-of-analysis. The conclusion summarizes the book's findings and offers some suggestion for improving the IDF's military thought by strengthening its institutionalized intellectualism. The Appendix compares Israeli and American military thought via a content analysis of *Maarachot* and *Military Review* pieces, prior to the end of the Cold War and in its aftermath.

Acknowledgements

I am grateful for the financial assistance of the Israel Science Foundation (ISF). I also wish to thank the anonymous referees of ISF for their valuable comments; Dima Course and Moran Weinreich for excellent research assistance; Elisheva Blusztejn for editing the book and for her valuable advice; Julian Deahl, Marcella Mulder, and Fem Eggers at Brill for making the process from book proposal to book so professional, smooth and efficient; the anonymous readers for their valuable suggestions; and the IDF's Staff and Command College and the National Defense College for inviting me to teach senior commanders for some fifteen years, which helped me get acquainted with their education process and sharpened my understanding of their way of thinking. My criticism of the IDF's military thinking comes from a place of empathy, appreciation, and concern. Last but not least, I would like to thank from the bottom of my heart to my wife Gila and my sons Ofer and Shahar for their love, encouragement and support.

List of Tables

1.1 Theory, doctrine, and planning 2
2.1 Military theory 1948-2008 in numbers and percentages 46
2.2 Military theory per period (percentages except for N) 47
2.3 Military theory (books) in numbers and percentages 49
2.4 Status per period (percentages except for N) 50
2.5 Military history 1948–2008 in numbers and percentages 60
2.6 Military history per period (percentages except for N) 60
2.7 Types of conflict per period (percentages except for N) 65
2.8 Ranks in 1948-2008 in numbers and percentages 69
2.9 Types of conflicts 1948-2008 in numbers and percentages 76
2.10 Levels of war 1948-2008 in numbers and percentages 77
2.11 Levels-of-war per period (percentages except for N) 78
2.12 Dimensions of strategy 1948-2008 in numbers and percentages 80
3.1 Types of conflict per period (percentages except for N) 87
A1 Status 1983–1987 149
A2 Status 1988–1994 149–150
A3 Status 1995–2000 150
A4 Status 2001–2004 150
A5 Status 2005–2008 150
A6 Rank 1983–1987 151
A7 Rank 1988–1994 152
A8 Rank 1995–2000 152
A9 Rank 2001–2004 153
A10 Rank 2005–2008 153
A11 Military theory 1983–1987 154
A12 Military theory 1988–1994 154
A13 Military theory 1995–2000 154
A14 Military theory 2001–2004 155
A15 Military theory 2005–2008 155
A16 Military history 1983–1987 155
A17 Military history 1988–1994 155
A18 Military history 1995–2000 156
A19 Military history 2001–2004 156
A20 Military history 2005–2008 156
A21 Types of conflict 1983–1987 156
A22 Types of conflict 1988–1994 157
A23 Types of conflict 1995–2000 157

A24 Types of conflict 2001–2004 157
A25 Types of conflict 2005–2008 158
A26 Dimensions of strategy 1983–1987 158
A27 Dimensions of strategy 1988–1994 158
A28 Dimensions of strategy 1995–2000 159
A29 Dimensions of strategy 2005–2008 159
A30 Levels of war 1983–1987 159
A31 Levels of war 1988–1994 160
A32 Levels of War 1995–2000 160
A33 Levels of War 2001–2004 160
A34 Levels of War 2005–2008 160

CHAPTER 1

Military Thought and the Formative Factors Affecting It

This chapter offers a research framework for the analysis of military thought. The chapter starts with a short presentation of two main indicators that are representative of the nature and quality of any military thought: its intellectual focus and modern focus. It then analyzes the formative factors affecting military thought, by pointing to the relationship between realist and non-realist – mainly cultural – formative factors, and by examining the influence of the formative factors at three levels of analysis – the systemic level, the unit (state) level, and the level of the individual.

The Intellectual and Modern Focus of Military Thought

Intellectual Focus

The intellectual focus of military thought can be assessed by the role played by history-based military theory within military thought, as compared to the role played by two other intellectual activities: doctrine and planning. Theory is universal, whereas doctrines and planning reflect responses for the particular, unique conditions under which each country and military operates. While doctrine refers to principles that are supposed to guide the effort to cope with the basic strategic conditions of the state and military, planning offers answers to the dynamic, ever-changing strategic environment. These distinctions are presented in the following two-by-two matrix.

But even within theory, a distinction is in place between theories that aspire to represent the entire spectrum of war and strategy, and the so-called middle-range theories, which focus on specific branches of strategy, such as air warfare, naval warfare, guerrilla warfare, mountainous terrain warfare, post-heroic warfare, etc.[1]

Theory is supposed to offer a toolbox for crystallizing doctrines and planning. But there are cases where the influence goes the other direction, where theory is shaped by doctrine and planning. For example, the US's decision to

1 On "middle-range theory," see Robert Merton, *Social Theory and Social Structure* (Illinois: The Free Press, 1949), pp. 3–12.

 | DOI 10.1163/9789004306868_002

TABLE 1.1 *Theory, doctrine, and planning*

	Universal	Unique
Basic	Theory	Doctrine
On-going		Planning

use nuclear weapons against Japan was a result of the recent availability of nuclear technology. Only at a later stage did the actual use of nuclear weapons during the war trigger the crystallization of nuclear doctrine for the US and a universal nuclear theory that were absent when the bombs were dropped on Hiroshima and Nagasaki. Another example pertains to the Strategic Air Bases study, which was a planning project conducted by RAND Corporation after World War II for the US Strategic Air Force. The project resulted, among other things, in the inception into strategic theory of theoretical concepts such as first and second strike capability.[2]

In light of the very practical nature of the military profession, the notion of intellectualism in this field is often met by suspicion, doubt, and resistance, and is described either as alien to practice or at least of limited relevance to it, as compared to experience, common sense and intuition. This is especially true as far as theory is concerned, despite the fact that a history-based military theory deserves to be considered the jewel in the crown of military thinking. "I never expect a soldier to think," declares a character in George Bernard Shaw's *The Devil's Disciple*.[3] Edward Luttwak, in a reference to the American army, complained about "the ignorance of the basics of the military art."[4] And Steven Knott criticized the tendency to prefer "doers" to intellectuals, and to deny any linkage between successes of talented commanders, such as Joshua Chamberlain or George Patton, on the battlefield and their intellectual skills.[5] An article in *Army* magazine complained about "anti-intellectualism in the

2 Albert J. Wohlstetter, "Economic and Strategic Considerations in Air Base Location: A Preliminary Review," <http://www.rand.org/about/history/wohlstetter/D1114/D1114.html>; Albert J. Wohlstetter, Fred S. Hoffman, R.J. Lutz, and Henry S. Rowen, "Selection and Use of Strategic Air Bases," <http://www.rand.org/pubs/reports/2006/R266.pdf>.

3 Act III of George Bernard Shaw's *The Devil's Disciple*.

4 Edward N. Luttwak, "On the Need to Reform American Strategy," in Philip S. Kronenberg (ed.), *Planning US Security: Defense Policy in the Eighties* (New York: Pergamon, 1982), p. 23.

5 Steven W. Knott, *Knowledge Must Become Capability: Institutional Intellectualism as an Agent for Military Transformation* (Carlisle Barracks: US Army War College, 2004).

American military,"[6] citing an American general who expressed discontent over the tendency to prefer the "muddy boots" soldier or "men of action" over "men of reflection" as a well-known characteristic of American military culture.[7]

This tension between theory and practice has already preoccupied classical military thinkers. Clausewitz was sure that "the powers of intellect" play a significant role in war, and bear practical dividends, especially in coping with the challenge posed by uncertainty.[8] He therefore recommended that commanders do not underestimate the intellectual aspect of their profession. But were Clausewitz to choose between knowledge and genius, it seems that he would have preferred an intuition-based "military genius," which "needs no theory."[9] In a moment of truth, he admitted that "in the art of war, experience counts for more than any amount of abstract truths."[10] A commander "need not be a learned historian nor a pundit, but he must [...] know the character, the habits of thought and action, and the special virtues and defects of the men whom he is to command [...]. This kind of knowledge cannot be forcibly produced by an apparatus of scientific formulas and mechanics; it can only be gained through a talent for judgment and by the application of accurate judgment to the observation of man and matter."[11]

Jomini, too, took intellectualism with a grain of salt. He believed that it is the combination of knowledge and skill that constitutes a superior commander and assures success. He did not expect commanders "to know a great deal but to know well; to know especially what relates to the mission appointed" to them.[12] "An ignorant man endowed with a natural genius [...] can do great things." However, "the same man, stuffed with false doctrines studied at school, and crammed with pedantic systems, will do nothing good unless he forgot what he had learned."[13] Although knowledge is best acquired via a combina-

6 Lloyd J. Matthews, "The Uniformed Intellectual and His Place in American Arms, Part I," *Army*, Vol. 52, No. 7 (July 2002), pp. 18–20; Lloyd J. Matthews, "The Uniformed Intellectual and His Place in American Arms, Part II," *Army*, Vol. 52, No. 8 (August 2002), pp. 31–40.

7 David W. Bamo, "Military Intellectualism," House Armed Services Subcommittee on Investigations and Oversight, 10 September 2009 <http://209.85.135.132/search?q=cache:XHd6OHgoT1wJ:armedservices.house.gov/pdfs/OI091009/Barno-Statement.pdf+military+intellectualism&cd=5&hl=en&ct=clnk>.

8 Carl von Clausewitz, *On War* (Princeton: Princeton University Press, 1976), p. 101.

9 Ibid., p. 145.

10 Ibid., p. 164.

11 Ibid., p. 186.

12 A.H. Jomini, *Summary of the Art of War* (Urbana: University of Illinois Press, 1947), p. 18.

13 Ibid., pp. 14–15.

tion of combat experience, good intuition, and formal learning, many commanders seem to prefer the first two channels, feeling uncomfortable with the third.

The skepticism regarding the practical value of intellectualism notwithstanding, there have been military figures, such as Comte de Guibert, Napoleon, Helmuth von Moltke, Johann Jacob August Ruehle von Lilienstern, Alfred von Schlieffen, Alfred Thayer Mahan, Ferdinand Foch, Basil H. Liddell Hart, John F.C. Fuller, Giulio Douhet, William (Billy) Mitchell, Mao Tse-tung, André Beaufre, Lucien Poirier, or Pierre Marie Gallois, and others, who have combined reading, thinking, writing and fighting, serving as a model of intellectual soldiers.

Modern Focus

The modern focus of military thinking can be evaluated by the extent to which it takes into consideration the entire spectrum of types of war and combinations thereof; relates to the entire pyramid of the levels-of-war, beyond the tactical and the strategic levels; acknowledges the central role played by the non-military dimensions of war and strategy; evaluates the relevance and the combination of material and non-material dimensions; and absorbs post-modern ideas and translates them into practice. More specifically, the modern focus is interested in the way military thought treats aspects such as:

- The unprecedented linkage between unconventional and sub-conventional war, e.g., terrorist and guerrilla groups possessing weapons of mass destruction;
- The blurred boundaries between conventional and sub-conventional war, e.g., the Hybrid War;
- The role played by nonstate players;
- The dominance of the material aspects of war and strategy in general and technology in particular;
- The role played by the non-military aspects of war and strategy;
- New types of relationship between the higher strategic echelons and the lower ones, as reflected by the simultaneous "strategization of tactics" and the "tacticization of strategy;"
- The growing importance of post-modern notions.

Realist vs. Cultural Formative Factors

Although security studies have a strong bias for realism, cultural aspects have emerged as a competing formative factor,[14] either in the form of an independent variable or as an intervening variable. The "cultures" most relevant to military thought are strategic culture and military culture.[15] A built-in tension exists between realist and cultural formative factors. It is only natural that realists would prefer structural, materialist explanations for players' behavior and thought to cultural, non-materialist, domestic ones, to which they attribute a very limited explanatory power, if at all, and a poor empirical record.[16] They would also reject the idea of relating to culture as an independent variable. At most, they might be ready to accept it either as an intervening variable or a dependent variable. Michael Desch, for example, argues that not only does realism outperform cultural explanations,[17] culturalists themselves are sometimes ambivalent about the degree of independent explanatory power cultural variables have in security studies. To this effect he cites Jeffrey Legro, a culturalist who admits that "reality can be socially constructed, but only with available materials and within existing structures. [...] When the contradiction between external conditions and cultural tendencies becomes too great, culture will likely adapt."[18]

Most culturalists, for their part, contend that culture does have an autonomous explanatory power.[19] John Duffield, for one, claims that a superior

14 For culture as context rather than causality, see Colin S. Gray, "Strategic Culture as Context: The First Generation of Theory Strikes Back," *Review of International Studies*, Vol. 25, No. 1 (January 1999), pp. 49–69.

15 Theo Farrell, "Culture and Military Power," *Review of International Studies*, Vol. 24, No. 3 (July 1998), pp. 407–16.

16 Michael C. Desch, "Correspondence," *International Security*, Vol. 24, No. 1 (Summer 1999), p. 174; Douglas Porch, "Military 'Culture' and the Fall of France in 1940: A Review Essay," *International Security*, Vol. 24, No. 4 (Spring 2000), pp. 157–80.

17 Desch, "Correspondence," p. 173; Michael C. Desch, "Culture Clash: Assessing the importance of Ideas in Security Studies," *International Security*, Vol. 23, No. 1 (Summer 1998), p. 170.

18 Jeffrey W. Legro. *Cooperation under Fire: Anglo-German Restraint during World War II*. (Ithaca: Cornell University Press, 1995), p. 231.

19 See, for example: Jeffrey W. Legro, "Military Culture and Inadvertent Escalation in World War II," *International Security*, Vol. 18, No. 4 (Spring 1994), p. 116; Elizabeth Kier, "Culture and French Military Doctrine Before World War II," in Katzenstein (ed.), *The Culture of National Security*, p. 187; Theo Farrell, "Isms and Schisms: Culturalism versus Realism in Security Studies: Correspondence," *International Security*, Vol. 24, No. 1 (Summer 1999), pp. 161–8; John S. Duffield, "Isms and Schisms: Culturalism versus Realism in Security

approach to understanding state behavior would start with cultural variables and invoke realism as a possible supplement only when the former are found wanting.[20] Explanations along these lines have been offered by Colin Gray for the US and the USSR during the Cold War;[21] by Alastair Johnston for China;[22] and by Thomas Berger for Germany and Japan.[23] Although this study deals with thinking rather than behavior, the arguments offered here seem to apply to thinking as well.

Reflective of the difficulty that arises in attributing military thought to either approach is the concept of "way of war," which characterizes the unique features of conducting war by different players. On the one hand, it could be argued that the way of war of a state or a military is shaped by the materialistic features of its strategic environment; for example, the weak players' preference of attrition warfare as a means of balancing the military and technological edge of their stronger opponents.[24] On the other hand, it has been suggested, either implicitly or explicitly, that Third World countries have a built-in cultural barrier to waging *Blitzkrieg*, which has channeled them to prefer attrition as their strategy.[25] The tension between realist and cultural explanations is also reflected in the "cult of technology," which will be addressed below.

Studies: Correspondence," *International Security*, Vol. 24, No. 1 (Summer 1999), pp. 172–80; Ted Hopf, "The Promise of Constructivism in IR Theory," *International Security*, Vol. 23, No. 1 (Summer 1998), pp. 171–200. Johnston perceives of culture as an ideational factor that could explain both realist and non-realist modes of thought and behavior, but he also acknowledges that the conclusions derived from his strategic cultural analysis of Maoist China's strategy could also have been predicted by structural realism. Alastair Iain Johnston, "Cultural Realism and Strategy in Maoist China," in Peter Katzenstein (ed.) *The Culture of National Security: Norms and Identity in World Politics* (New York: Columbia University Press, 1996), pp. 216–68.

20 Duffield, "Isms and Schisms."

21 Colin Gray, "National Styles in Strategy: The American Example," *International Security*, Vol. 6, No. 2 (Fall 1981); Colin S. Gray, *Nuclear Strategy and National Style* (London: Hamilton, 1986), pp. 36–7.

22 Johnston, "Cultural Realism and Strategy in Maoist China."

23 Thomas U. Berger, "Norms, Identity, and National Security in Germany and Japan," in Katzenstein (ed.), *The Culture of National Security*, pp. 317–56.

24 Avi Kober, *Israel's Wars of Attrition* (New York: Routledge, 2009).

25 Anthony H, Cordesman and Abraham Wagner, *The Lessons of Modern War*, Vol. 2: *The Iran-Iraq War* (Boulder: Westview, 1990), Chs 12 and 15; R.M. Cassidy, *Russia in Afghanistan and Chechnya: Military Strategic Culture and the Paradoxes of Asymmetric Conflict* (Carlisle: US Army War College, Strategic Studies Institute, 2003), p. 53. For the difficulty in alluding culturally distinct strategic behavior to Third World countries and armies see, for example, Joseph Rothschild, "Culture and War," in Stephanie G. Neuman and Robert E.

Is There Such Thing as a Universal Strategic Culture or Military Culture?

Thinkers are divided on this question, too. Some thinkers hold that culture is unique.[26] To cite Colin Gray, it is about "persisting socially-transmitted ideas, attitudes, traditions, habits of mind and preferred methods of operation specific to a particular geographically-based security community that has had a unique historical experience."[27] Dima Adamsky discussed the differences in the adoption and implementation of RMA between Soviets, Americans and Israelis as a result of different cultures. The Soviets were the first to identify a basic change in the nature of war as a result of a technological breakthrough, thanks to their intellectualism; the Americans were the first to offer the revolutionary technology, lacking the dialectical and holistic thinking required for fully understanding the changes that took place in the nature of war; and the Israelis, who were intellectually weak but good at improvisation and practice, were the first to use these technologies on the battlefield.[28] Eitan Shamir showed how the attitudes of militaries to the doctrine of mission command were affected by cultural differences between the American, British and Israeli militaries.[29] Stephen Rosen pointed to the existence of unique sub-cultures in the American military, with the US Army, the Marines or the Navy having a distinct way of thinking about war.[30] Another example hails from the 19th and 20th centuries: Whereas Prussian military thinkers were educated in a militaristic tradition that accepted war as a fact of life, British military thinkers during the interwar period (including John F.C. Fuller, whose commitment to democratic values was far less scrupulous) were inspired by a liberal-democratic tradition. As such, they perceived of war as something evil, and emphasized the importance of enhancing peace rather than achieving victory

Harkavy (eds), *The Lessons of Recent Wars in the Third World*, Vol. 2 (Lexington: Lexington Books, 1987), p. 53; Norwell De Atkine, "Why Arabs Lose Wars?" *Middle East Quarterly*, Vol. 6, No, 4 (1999), pp. 17–27; Kenneth M. Pollack, *Arabs at War: Military Effectiveness 1948–1991* (Lincoln: University of Nebraska Press, 2002).

26 See for example Elisabeth Kier, *Imagining War: French and British Military Doctrine between the Wars* (Princeton: Princeton University Press, 1999), p. 144; Colin S. Gray, *Modern Strategy* (Oxford: Oxford University Press, 1999), p. 131. See also Lawrence Sondhaus, *Strategic Culture and Ways of War* (London: Routledge, 2006).

27 Gray, *Modern Strategy*, p. 131.

28 Dima Adamsky, *The Culture of Military Innovation* (Stanford: Stanford University Press, 2010), pp. 4, 134.

29 Eitan Shamir, *Transforming Command: The Pursuit of Mission Command in the US, British, and Israeli Armies* (Stanford: Stanford University Press, 2011).

30 Stephen P. Rosen, *Winning the Next War* (Ithaca: Cornell University Press, 1991), p. 19.

on the battlefield. Unlike the Prussian school that did not question the King's prerogative in defining the collective good in general and the war objectives in particular at his discretion, the liberal-democratic school stressed not only the notion of just war as the only legitimate war, but also the need for an elected political echelon as a guarantee for the realization of the people's interests and will, otherwise it would be replaced in free elections, something that would not happen in authoritarian regimes.[31]

A different view was suggested by Samuel Huntington, Morris Janowitz and Barry Posen, who identified features that were common to every military organization, such as the tendency to stick to traditional conceptions or conservative biases,[32] which made it difficult for them to adapt to a new reality of war – a phenomenon also known as entrenched traditionalism.[33]

An in-between opinion is held by Alastair Johnston, Steven Ott, John Duffield and Theo Farrell. According to Johnston, different players or organizations may have similar cultures;[34] Ott suggests that there are both common and unique threads among organizations' cultures, which reflect a complex interplay of similarities and differences;[35] Duffield believes that cultural theories are not inherently limited to emphasizing the uniqueness of cases;[36] and Farrell argues that the boundary between world culture and local culture has

31 On German militarism, see Gerhard Ritter, *The Sword and the Scepter: The Problem of Militarism in Germany* (Coral Cables: University of Miami Press, 1969–1973); Azar Gat, *Policy and War in Modern Military Thought* (Tel Aviv: Maarachot, 1984) [Hebrew], p. 103. On the British approach, see Liddell Hart, *Thoughts on War*, p. 42; John F.C. Fuller, *The Conduct of War 1789–1961* (New Jersey: Rutgers University Press, 1961), pp. 59–76.

32 Samuel P. Huntington, *The Soldier and the State: The Theory and Politics of Civil-Military Relations* (Cambridge: Belknap Press, 1957); Morris Janowitz. *The Professional Soldier: A Social and Political Portrait* (New York: The Free Press, 1960), p. 13; Barry Posen, *The Sources of Military Doctrine: France, Britain, and Germany Between the World Wars* (Ithaca: Cornell University Press, 1984).

33 Basil H. Liddell Hart, *Thoughts on War* (London: Faber & Faber, 1943), p. 30; Terry Terriff and Theo Farrell, "Military Change in the New Millennium," in Farrell and Terriff (eds), *The Sources of Military Change*, p. 265. R.A. Mason, "Innovation and the Military Mind." Online. Available at <http://www.au.af.mil/au/awc/awcgate/au24-196.htm>.

34 Alastair Iain Johnston, "Thinking about Strategic Culture," *International Security*, Vol. 19, No. 4, (Spring 1995), pp. 32–64.

35 Steven J. Ott, *The Organizational Culture Perspective* (Pacific Grove, CA: Brooks/Cole, 1989, pp. 74–84.

36 Duffield, "Isms and Schisms."

become blurred, and norms may have multiple residences at the world and local levels.[37]

Formative Factors: The Systemic Level

Clausewitz rightly characterized war as "more than a true chameleon that slightly adapts its characteristics to the given case."[38] This section focuses on changing materialistic and non-materialistic systemic conditions and developments in war, and their impact on military thinking.

Technology as a Dominant Dimension

Technology is the most dynamic dimension in modern war. Technological changes in modern time differ from those in classical antiquity. As J.E. Lendon points out, "There were some slight technological advances in ancient land warfare [...] but none was so emphatic an improvement that all nations were obliged to adopt it or fight at a severe disadvantage, as the situation created by modernity. [...] All in all, there were far less technological changes in the eight hundred years from 500 BC to 400 AD than in the forty years from 1910 to 1950; far less technological advance in any ancient century than in one year of either of the World Wars."[39]

The impact of technology on military thought has many aspects, such as the squaring of Clausewitz's triangle so that it would include the material dimension, in which technology plays the central role; the addition of technology to Michael Howard's non-operational dimensions of strategy; the emergence of the operational level-of-war, between tactics and strategy; the creation of the third (aerial) dimension and air warfare, space warfare, and cyber warfare; the notion of the deep battle (see below); technology as the facilitator of post-heroic warfare (due to the emergence of precision weapons, unmanned platforms etc. and the resulting dramatic decrease of casualty rates in war); the linkage between the glut of information available to commanders and the eternal element of friction; the emergence of network-centric warfare and its impact on the command system; etc.

---

37 Theo Farrell, "Isms and Schisms: Culturalism versus Realism in Security Studies: Correspondence," *International Security*, Vol. 24, No. 1 (Summer 1999), pp. 161–8.

38 Clausewitz, *On War*, p. 89.

39 J.E. Lendon, *Soldiers and Ghosts: A History of Battle in Classical Antiquity* (New Haven: Yale University Press, 2005), pp. 8–9.

Technology has also played a major role in creating revolutions in war that military thought had both to explain and cope with their consequences. More specifically, the marriage of deep technological and societal changes, i.e., the emergence of mass armies and the technological revolution in the early 19th century, brought about a revolution in war that has been reflected in military thought. The nuclear revolution was a result of the availability of technology that could destroy entire societies; and cyber warfare harnesses technology to topple the enemy's society, economy and army, and has at least the potential of fighting without using traditional force.

The marriage of the technological and the operational dimensions of war and strategy, on the other hand, has accounted for significant changes in war and strategy but can hardly be considered a revolution, even if it was initially referred to as such. This is true for the so-called Military-Technical Revolution (MTR) or the more recent Revolution in Military Affairs (RMA).

The dominant role played by technology has also left its imprint in the form of the cult of technology, which is mostly typical of developed Western democracies. In his general work on the rise of technology in Western culture, Neil Postman argued that towards the 21st century, human society has transformed from technology user to technology dominated.[40] This diagnosis seems to apply to the military field, as well. And indeed, according to Fuller, "tools or weapons [...] form 99 percent of victory. [...] Strategy, command, leadership, courage, discipline, supply, organization and all the moral and physical paraphernalia of war are nothing to a high superiority of weapons – at most they go to form the one percent which makes the whole possible."[41] For technology fans, the non-material components of military buildup, such as professional education, training or doctrines are less identifiable and quantifiable and often take longer periods of time to manifest themselves than technology.[42] For them, strategy equals targeting, and success is measured by the number and nature of targets destroyed.[43]

A criticism of the Pentagon's belief in the RMA called the over-belief in technology "blind faith," "the Maginot Line of the 21st century" and "fairy-tale."[44]

---

40 Niel Postman, *Technopoly: The Surrender of Culture to Technology* (New York: Vintage Books, 1993).

41 John F.C. Fuller, *Armament and History* (New York: Scribner, 1945), p. 18.

42 Meir Finkel, "The Cult of Technology in the IDF," *Maarachot* 407 (June 2006), pp. 40–45; Handel, *Masters of War*, p. 85.

43 Grant T. Hammond, *The Mind of War: John Boyd and American Security* (Washington, Smithsonian Books, 2001), p. 207.

44 John A. Gentry, "Doomed to Fail: America's Blind Faith in Military Technology," *Parameters*, Vol. 22, No. 4 (Winter 2002–03), pp. 88–103.

Works by Theodore Roszak on the "information glut;"[45] James Gibson on the "technowar" in Vietnam and its aftermath;[46] Williamson Murray on the allure of technology in America in the post-Vietnam period; Robin Higham on the "American interest in things mechanical;"[47] or Frank Hoffman on the American tendency to worship at the altar of technology and to think of conflicts as a mechanistic engineering exercise[48] – are all very critical of the cult of technology. Finally, Michael Handel argued that technology is often ineffective in asymmetrical conflicts, in which the nonstate player chooses to use insurgency, or in symmetrical confrontation in which the enemy develops and applies technological counter-measures. It is also difficult to guarantee absolute information dominance, to overcome friction, or to translate military achievements to political gains.[49]

The Broadening of War

A particularly notable manifestation of the dynamic nature of war at the systemic level has been its broadening since the 17th and 18th centuries. Below is an account of the process and some of its expressions.

The Emergence of the Large Citizens' Armies, and War Withdrawing from the Direct Battlefield

Whereas the skirmishes between knights during the Middle Ages, the mercenary wars during the 16th century, the trade wars during the 17th and 18th centuries, or the wars conducted by professionals during the 18th century[50] had not really involved the society at large, and consequently wars had been less violent and brutal, and more rational, controlled and limited in scope, during the 19th century things changed significantly. Wars transformed into multi-dimensional struggles across national borders, and often involved the civilian rear. The confrontation between armies became just one facet of armed conflicts, though the pivotal one.

---

45 Theodore Roszak, *The Cult of Information: A Neo-Luddite Treatise on High-Tech, Artificial Intelligence, and the True Art of Thinking* (Berkeley: University of California Press, 1994).

46 James W. Gibson, *The Perfect War: Technowar in Vietnam* (New York: Atlantic Monthly Press, 1986).

47 Robin Higham, *The Military Intellectuals in Britain: 1918–1939* (New Brunswick: Rutgers University Press. 1966), p. 12.

48 Frank G. Hoffman, "The Anatomy of the Long War's Failings," *FPRI's Newsletter*, Vol. 14 No. 16 (May 2009). Available at <http://www.fpri.org/footnotes/1416.200905.hoffman.longwarsfailings.html#note5>.

49 Handel, *Masters of War*, pp. xx-xxiv.

50 Michael Howard, *War in European* History (London: Oxford University Press, 1976).

The watershed in this respect was of course the French Revolution and the Napoleonic wars that changed the nature of war radically and gave it a far more total character. The societal dimension of war pushed war towards extremity, a trend that was arrested only by intense traumas, such as the destruction caused by World War I, or the use of nuclear weapons at the end of World War II. The involvement of the population in the conduct of war deepened even further with the emergence of the European parliaments and the intensification of popular interest in state affairs.

Technology, whose dominant role was discussed above, played a major role in the process. It had started affecting it even before the emergence of mass armies, as a result of the proliferation of firearms, and the growth of the infantry at the expense of the cavalry in the 16th century. But in the wake of the Industrial Revolution, which took place in the 1830s, its influence on the broadening of war has increased significantly. For example, the invention of the railway in the 19th century made it possible to expedite large forces to the battlefield with relative speed and keep them supplied, which reinforced the national aspect of military conflict even further; technological changes have deepened the battlefield and have created a linkage between the battlefield and the civilian rear, in addition to introducing a third dimension – aerial warfare – and later on space warfare and cyber warfare; technological changes have improved the command and control of large formations; and have deepened the public's involvement in the war by enabling it, through the mass media, to remain informed about events on the battlefield, in real or near-real time, and by turning society into a direct target.

"National Logistics"

The aforementioned changes gave new meaning to logistics. Until the 19th century logistics had referred to efforts made to keep the fighting men supplied with food; but as the military leadership became more aware of the multiple tasks involved in the conduct of operations, logistics took on a far more comprehensive meaning, involving (according to Jomini) all functions and tasks that were not purely operational – in other words, not only food supplies, but also manpower, communications, engineering, and so forth.[51] Clausewitz belittled the value of logistics, believing that it belonged to activities that took place mainly before the outbreak of the war, as such being rather an esoteric element to the art of war.[52] The opposite opinion became widely held in the

---

51 A.H. Jomini, *Summary of the Art of War* (Urbana: University of Illinois Press, 1947), pp. 132–5

52 Clausewitz, *On War*, pp. 128–9.

late 19th and 20th century: under the influence of wars such as the American Civil War and the two World Wars, war tended to be regarded as one gigantic logistical operation. Alfred Thayer Mahan linked the efficiency of military operations and the degree of success in rallying the economic, industrial and other resources of the nation in support of the army. In so doing, he raised logistics to the level of grand-strategy.[53] This trend continued into the 20th century, where it found expression in the works of Fuller and Liddell Hart. The more intensive war became, the more it was accompanied by large-scale logistic operations, whereas the capacity to continue a war over a long period of time has come to depend more and more upon the capacity to translate national power in general, and industrial infrastructure in particular, into a military buildup, either through local production or procurement from outside parties.

The transition from personal weaponry to weapon systems at the beginning of the 20th century strengthened the "logistic tail" relative to the "operational teeth." However, with the increased emphasis on firepower during the second half of the 20th century at the expense of maneuver, logistical systems in highly technological armies have focused on supporting firepower, and became more centralized, which threatened to cripple the combat units' logistical autonomy and went against strategic logic, which is different from non-military logic.

Expressions of the Broadening of War in Military Thought

The broadening of war has had an impact on the components of the modern focus of military thought.

The Dimensions of War: Squaring the Triangle

With the growing importance of technology in war and strategy, Clausewitz's trinity – the triangle consisting of government, people and military – which had been identified with modern war, has become partially obsolete as it has consisted of non-material dimensions. With the growing importance of the material aspect of war, such as technology and economy, thinkers have felt that the time has come to "square the triangle" and to include particularly the technological dimension.[54]

---

53 Alfred Thayer Mahan, *The Influence of Sea Power Upon History, 1660–1783* (Boston: Little, Brown, 1940).

54 Michael Handel, "Clausewitz in the Age of Technology," in Michael Handel (ed.), *Clausewitz and Modern Strategy* (London: Frank Cass, 1986), pp. 51–92.

At the same time, the people (the society) has become not merely a resource mobilized for supporting the war effort on the direct battlefield (as reflected in Liddell Hart's initial notion of grand-strategy), but also a major target to hit. This has been manifested in the military thought of air power thinkers Douhet, Seversky and Mitchell, and guerrilla and terror strategists.

Also, in modern war the political echelon (the government) and the military echelon have become decentralized. States are usually ruled by a government rather than a single ruler, and general staffs comprise of a group of individuals, each functioning as an expert on a different aspect of military buildup or operations.

Dimensions of Strategy

According to Michael Howard, the predominant – if not virtually exclusive – status of the operational dimension of strategy has been eroded by the growing influence of the societal and technological dimensions, as well as by the rapid expansion of the logistical dimension, to the point where the latter has become a central factor in modern military strategy.[55] In our era it has become impossible to mention strategy without taking into account all four dimensions, even though their relative significance may vary from one context to another. Anyway, the above process has created the need to inculcate in military thought non-military dimensions and their interaction with the military ones.

Technology has contributed to the broadening of war by increasing range and destructive power of weapon systems, which have created the notion of deep operations. Deep operations have initially been referred to in military thought in the framework of maneuver, e.g., the Soviet concept of mobile, combined operations, breaking through the enemy's defenses and moving into its strategic depth in the 1920s,[56] but have during the early 1980s taken on the form of a combination of firepower and maneuver, e.g., the American concept of AirLand operations.[57] At the same time, technology has had an impact on the narrowing of war, but this will be referred in the following section.

---

55 Michael Howard, "The Forgotten Dimensions of Strategy," *Foreign Affairs*, Vol. 57, No. 5 (Summer 1979), pp. 975–86.

56 Richard Simpkin, *Deep Battle: The Brainchild of Marshal Tukhachevsky* (London: Brassey's, 1987).

57 R. Kent Laughbaum, *Synchronizing Airpower and Firepower in the Deep Battle* (Maxwell Air Force Base: Air University Press, January 1999 <http://www.dtic.mil/cgi-bin/GetTRDoc?AD=ADA392508>.

The Levels-of-War

Modernity is also associated with the expanding of levels-of-war from a two-level pyramid until the late 18th century-early 19th century, which had consisted of strategy and tactics, to four, with the two additional levels of the operational level and the grand-strategic level.

B.H. Liddell Hart's concept of grand-strategy was initially crystalized against the backdrop of the strong linkage between war efforts in the rear and military operations, with the former expected to provide the war effort on the front with the necessary resources. Liddell Hart realized that war had spilled beyond the battlefield, that it had become a confrontation between national powers, or between equations of national security, and that as a result military victory on the battlefield had, to a large extent, become dependent on the resources that the nation was able and willing to devote to the conduct of war on the battlefield. Herewith, more or less, the conceptual basis was laid for the concept of national security. This approach focuses on the interaction between political, societal, economic, demographic and technological dimensions, on the one hand, and the military dimension, on the other, with the latter remaining the central one. There has been a sting in the tail, though, since the focus on grand-strategy, that is, on the national level, has detracted from the preoccupation with the theoretical aspects of military strategy. With the civilian rear becoming a major target for adversaries, grand-strategy has been given an additional meaning. This is reflected in the works of airpower thinkers Giulio Douhet, Alexander Seversky, Billy Mitchell, and the guerrilla and terror strategists.

Another impact of the broadening of war on the levels-of-war has been the appearance of the operational level, as yet another new level. This level became necessary once it was acknowledged that maneuvers and stratagems, applied above the tactical level, could affect the way battles were conducted and their chances for victory. An important factor that made this feasible was technological developments, such as the introduction of the train, and the mechanization of the battlefield.

As early as the 18th century, Guibert distinguished between elementary tactics, which referred to the structure and organization of military units and the technique of their operation, and grand tactics (what we would call nowadays the operational level), dealing with the plans, decision-making and activities that influence the outcome of a military confrontation before it has begun or during the confrontation.[58] The idea that each level-of-war can be affected by the levels above it received yet further impetus as a result of the vastly increased

58 Palmer, "Frederick the Great, Guibert, Bülow...," p. 107.

influence on military thinking of military geniuses such as the generals of the Thirty-Year War or Frederick the Great during the 18th century Seven-Year War, or Napoleon at the beginning of the 19th century. But there still remained considerable skepticism about the impact of higher levels on the lower ones, a skepticism that is clearly reflected in Clausewitz's thought, which focuses on the conduct of war on the battlefield itself while minimizing the value of plans and actions taken prior to the fighting, such as maneuvers, on the actual outcome of the confrontation. Generally speaking, Clausewitz tended to regard strategy as merely a level that renders significance to moves at lower levels. He considered strategic success to be the result of incremental military successes at the lower levels, in a bottom-up process.[59]

The strengthening of the non-military aspects of war and of the higher levels-of-war in the post-World War I era notwithstanding, the tactical level is still very popular with war-related writings by the military. This is also evident in the *Maarachot* survey. The explanation seems to be twofold: it is easier to think narrowly, and commanders seem to be mostly affected by their tactical experience and training.

Furthermore, a direct connection has often occurred, particularly in LICs and wars of attrition, between the two extremes of the levels-of-war pyramid. Reflective of this process have been the so-called tacticization of grand-strategy and the parallel but opposite direction grand- strategization of tactics, for reasons that will be discussed later in this chapter.

Mission Command

The appearance of Defense Ministries and modern General Staffs not only reflected the growing scope and complexity of strategy and security, but also offered solutions to all kinds of problems connected with the management and control of large popular armies that existed from the 19th century onwards. The process raised questions about centralized versus decentralized command systems. In contrast to the centralized command of Napoleon, who personally supervised his huge army without any visible command and control system in the modern sense, nor by adopting the mission-command system, most members of the German school, from Gerhard von Scharnhorst, August von Gneisenau and Clausewitz, via Friedrich von Bernhardi and Helmuth von Moltke ("the Elder"), up to the commanders of the German army between the two World Wars, did in fact adopt the mission command. The underlying assumption of this approach, which granted greater autonomy and a considerable degree of discretion to the lower command echelons, was the existence of

59 Clausewitz, *On War*, p. 363.

friction on the battlefield and the difficulty of implementing centralized control of large armed forces.[60] The mission command assigned the commanders an objective or mission, a zone of operations, and a timetable, but left the method used for accomplishing the mission to each commander's discretion. A good decentralized mission-oriented command was not supposed to be based on improvisation but rather on a thorough educational and training process during which all commanders acquired the same set of professional tools, which they would be later be using according the specific conditions on the particular battlefield each of them would be operating on. In the Prussian army, first under Moltke and then after his successors, superiors had confidence in their subordinates, based on the assumption that having acquired a similar set of theoretical and doctrinal tools, each of them would act exactly as his superior would have acted had he been in his place.

The Narrowing of War

The broadening of war has been accompanied by developments that have decreased the size of military formations as a result of weapons' enhanced lethality, improved performance, and rising cost; lowered the casualty rates as a result of weapons' accuracy; and tightened the political control over military operations.

The Impact of Weapons' Lethality and Improved Performance

The more lethal weapon systems have become, the smaller the numbers needed for producing the desired effect.[61] Effectiveness has improved so dramatically that an infantry battalion of today possesses the firepower of a division during World War II, and two F-117 squadrons' simultaneous attack is sufficient for inflicting the same damage that required thousands of bombers attacking sequentially back in World War II.[62]

This tendency was reflected in the number of US troops deployed on the battlefield in the HIC chapter of the 2003 Iraq War, which was half the number

60 On the rationale behind the mission command, see for example Walter Goerlitz, *History of the German General Staff 1657–1945* (New York: Praeger, 1957), pp. 41, 75–6.

61 Henry S. Rowen, *Intelligent Weapons: Implications for Offense and Defense* (Tel-Aviv: Jaffee Center for Strategic Studies, 1988), p. 19; Lance Davidson, "The Impact of Precision Guided Munitions on War," *Brassey's Yearbook 1984* (London: Brassey, 1984); Eliot A. Cohen, "A Revolution in Warfare," *Foreign Affairs*, Vol. 75, No. 2 (March/April 1996), p. 45.

62 Ofer Shelah, *The Israeli Army: A Radical Proposal* (Or Yehuda: Kinneret, Zmora-Bitan, Dvir, 2003) [Hebrew], p. 42.

of those deployed in the 1991 Gulf War (250,000 and 500,000, respectively).[63] The overwhelming heavy-armored offensive across a broad front of the 1991 Gulf War (often alluded to as the "Powell Doctrine") was replaced in 2003 by a smaller, faster-moving attack force on the ground, capable of launching precision weapons at vast distances and taking advantage of sophisticated reconnaissance systems, combined with overwhelming air power.[64] The reduced number of troops, however, proved to be insufficient for effective control of the country during the LIC chapter of the war (see discussion of the LIC aspect, below).

The Rising Cost of Weapons Systems

The high cost of precision weapons compared to iron bombs notwithstanding, target destruction cost would often be cheaper with precision munitions. For example, classical bombing missions require far more sorties and might involve heavy support by escort aircraft, electronic warfare aircraft, airborne refueling and intelligence aircraft.[65] On the other hand, the cost of military technology in the developed industrial states has exceeded the general inflation rate by some 3–5 percent.[66] Furthermore, according to "Augustine's Laws," while defense budgets grow linearly the unit cost of a new military aircraft grows exponentially. This might mean that in the year 2054, the entire US defense budget will purchase just one tactical aircraft.[67] This encourages attempts to temper the growing cost of sophistication by efforts to reduce one's forces, if possible.

Lower Casualty Rates

The combination of unmanned platforms, improved accuracy and enhanced range of weapons, and Western democracies' casualty aversion in cases where no vital interests are involved is the major explanation for the drastic decline in casualties in the 21st century. Whereas the Vietnam War claimed the lives of at least 4 percent of the Vietnamese population, in the 1991 Gulf War only some 0.07 percent of the Iraqi population was killed. In Kosovo the percentage

63 Max Boot, "The New American Way of War," *Foreign Affairs* 82, no. 4 (July/August 2003), p. 43.

64 Seymour M. Hersh, "Offense and Defense," *The New Yorker*, 7 April 2003.

65 Zeev Bonen, "Technology in War: Preliminary Lessons from the Gulf War," in JCSS Study Group, *War in the Gulf: Implications for Israel* (Tel Aviv: Jaffee Center for Strategic Studies, 1992), p. 171.

66 Yaacov Lifshitz, "Managing Security Resources in the 2000s," in Haggai Golan (ed.), *Israel's Security Web* (Tel-Aviv: Maarachot, 2001) [Hebrew], p. 51.

67 Norman Ralph Augustine, *Augustine's Laws* (New York: Viking, 1983), Law No. 16, 1984.

declined to 0.005 percent, while in Afghanistan and Iraq the numbers rose slightly to 0.01–0.03 percent of the population of the respective countries.[68] This happened contrary to the expectation that after 9/11 the US would be less averse to killing enemy civilians.

The sharp drop in civilian casualty rates demonstrates that it was the 1991 Gulf War rather than the 2003 Iraq War that constituted a watershed in this respect. Since the Gulf War, the US has managed to leave intact the lion's share of civilians on the enemy's side. In an almost surrealistic report from the Iraqi capital during the 2003 Iraq War, Charles Krauthammer described "plumes of smoke from precision strikes on Saddam's instruments of power while the city lights remained on and cars casually traversed the streets."[69]

Control of War

The existence of weapons of mass destruction, the sensitivities and vulnerabilities of Western democracies involved in LICs, and the unprecedented availability of effective information sources and means of command and control at the political leadership's disposal require and enable tight control over the use of force. According to Eliot Cohen et al., the more force applied in counterinsurgency operations, the greater the chance of collateral damage, mistakes, and enemy propaganda portraying the stronger side's military activities as brutal (what is often referred to as excessive use of force).[70] The political echelon therefore often finds itself directly interfering in tactical matters and/or restricting operations or battles, bypassing the strategic and operational levels (the "tacticization of strategy").

68 The percentage calculus is based on the following numbers: Gulf War: 13,000 out of 19,000,000 Iraqis; Kosovo: 500 out of 10,000,000 Serbs; Afghanistan: 3,500 out of 28,000,000 Afghanis; Iraq War: 3,750 out of 26,000,000. Karl Schoenberger, "Civilian Casualties will Erode Support, Inflame Hatred Among Iraqis," *San Jose Mercury News*, 26 March 2003; Peter Ford, "Surveys Pointing to High Civilian Death Toll in Iraq," *Christian Science Monitor*, 22 May 2003; *Reuters*, 9 July 2003; Carl Conetta, "The Wages of War: Iraqi Combatant and Noncombatant Fatalities in the 2003 Conflict," Project on Defense Alternatives *Research Monograph* #8, 20 October 2003; <http://www.cbc.ca/news/iraq/issues_analysis/casualties_postiraqwar.html>; <http://www.comw.org/pda/0310rm8.html>.

69 Charles Krauthammer, "Gulf War II is First of its Kind," *TownHall.com*, 10 April 2003.

70 Eliot Cohen, Conrad Crane, Jan Horvath, and John Nagl, "Principles, Imperatives, and Paradoxes of Counterinsurgency," <http://usacac.army.mil/CAC/milreview/English/MarApr06/Cohen.pdf>.

Have LICs contributed to the narrowing of war?
At least by name, LICs imply conflicts that are less intensive than HICs, which apparently could be presented as a narrowing factor. Martin Van Creveld portrays an atomization process that will take us back to the pre-state era, with smaller, non-state entities fighting each other with low tech weapons, on relatively much minor issues and with much less destructive results.[71] He criticizes Stephen Biddle's *Military Power*[72] for ignoring the fact that major conventional war between or against major military powers has been shrinking to the vanishing point and for misrepresenting the conflicts in Bosnia, Croatia, Eritrea, Zaire, Rwanda and Azerbaijan as mid- to high-level conflicts.[73]

But do LICs really deserve to be considered expressions of the narrowing of war? The answer seems to be in the negative. For too many people, LICs have been high-intensity rather than low-intensity conflicts,[74] as more people have been killed in LICs, or "small wars" than died in World War II.[75] During the Cold War, millions of casualties were inflicted on peoples entangled in LICs.[76] In the period of 1975–94, for the first time, intrastate war deaths exceeded the interstate war deaths – and by a wide margin.[77] While civilians accounted for

71 Martin Van Creveld, *The Transformation of War* (New York: The Free Press, 1991), pp. IX, 225.

72 Martin Van Creveld, "Less than Meets the Eye," *Journal of Strategic Studies*, Vol. 28, No. 3 (June 2005), p. 450; Cassidy, *Russia in Afghanistan and Chechnya.*

73 Van Creveld, "Less than Meets the Eye," pp. 449–52.

74 As one researcher put it, "for those unfortunate enough to be involved in the suffering caused by insurgency or chronic terrorism, the phrase low-intensity conflict does not begin to capture the trauma and tragedy of their lives. [...] As one might expect, the phrase does not enjoy similar popularity in Afghanistan, or Angola, or El Salvador, or Lebanon, or anywhere else that war is tangible reality". Loren B. Thompson, "Low-Intensity Conflict: An Overview", in Loren B. Thompson (ed.), *Low-Intensity Conflict: The Pattern of Warfare in the Modern World* (Lexington: Lexington Books, 1989), pp. 1–2. The devastating effect of Vietnam was summarized by Eliot Cohen in the following words: "The war in Vietnam, for example, killed 60,000 Americans, bred turmoil in the US society, devastated Vietnam for a generation, left Laos a backwater and was at least indirectly responsible, in neighboring Cambodia, for one of the greatest horrors of the twentieth century", Eliot A. Cohen, "The 'Major' Consequences of War", *Survival*, Vol. 41, No. 2 (Summer 1999), p. 143.

75 William J. Olson, "Preface: Small Wars Considered," *Annals of the American Academy of Political and Social Studies*, Vol. 541 (September 1995), p. 8.

76 Ruth L. Sivard (ed.), *World Military and Social Expenditures* (Washington, DC: World Priorities, 1987), pp. 29–31.

77 Peter Wallensteen and Margareta Sollenberg, "The End of International War? Armed Conflict 1989–95," *Journal of Peace Research*, Vol. 33, No. 3 (August 1996), p. 356.

10 percent of those killed during World War I, and 52 percent of those killed in World War II, they make up some 90 percent of contemporary war deaths.[78]

LICs have also brought about the displacement of millions of people. The world total of refugees grew from around two million in 1970 to over 16 million in 1995, with 20 to 30 million people displaced within their own national borders.[79] In terms of duration, many LICs last more than 10 years. Furthermore, a sustained attempt to wear the enemy down, which is central to LICs, cannot be confined to operations on the direct battlefield. It must spill over to the grand-strategic level, where it is aimed at the enemy's economy and society, with the civilian rear becoming the center of gravity.

Expression in Military Thought

A strong feeling has spread among Western countries and militaries that war has become less costly in terms of casualties and money, and that a rapid and relatively bloodless victory, achieved by small and smart armies, is now at hand.[80] Two factors are mentioned in the military literature with regard to symmetrical and asymmetrical conflicts and the military's size, which challenge the notion of small and smart militaries conducting quick, costless wars – the *troop density paradox*, and the *force-to-space ratio*.

According to the troop density paradox, even in an RMA era, LIC challenges may require more troops than HIC ones. Since the mid-1990s, history-based theories regarding troop density in "stability operations," which fall under the category of LICs, have been offered by military analysts. Most density recommendations have fallen within a range of 20–25 soldiers per 1,000 residents in an area of operations. According to RAND analyst James Quinlivan, such ratios existed in Bosnia, 1996 (22.6 troops per 1,000 residents) and in Kosovo, 1999 (23.7 to 1,000).[81] In other conflicts, on the other hand, such as Somalia (1993), Haiti (1994), Afghanistan (2002), and Iraq, the ratios were well under 20 troops

78 David Turton, "Introduction", in David Turton (ed.), *War and Ethnicity: Global Connections and Local Violence* (San Marino: University of Rochester Press, p. 3.

79 Ibid., p. 2. Prominent examples include: Afghanistan (as of 1978): 5.2 million people; Angola (1975-): 2 million; Azerbaijan (1990-): 1.7 million; Bosnia (1992-): 2.5 million; Liberia (1989-): 1.7 million; Rwanda (1990-): 2 million; Somalia (1990-): 1 million; Sri Lanka (1983-): 1.2 million. Michael Brown, "Introduction", in Brown (ed.), *The International Dimensions of Internal Conflict*, pp. 4–7.

80 Andrew P.N. Erdmann, "The US Presumption of Quick, Costless Wars," in John F. Lehman, and Harvey Sicherman (eds), *America the Vulnerable* (Philadelphia: Foreign Policy Research Institute, 2002), pp. 48–72.

81 James Quinlivan, "Burden of Victory: The Painful Arithmetic of Stability Operations," *RAND Review*, Vol. 27, No. 3 (Summer 2003), available at <http://rand.org/pubs/corporate_

per 1,000 residents – 4.6 in Somalia, 3.5 in Haiti, 0.5 in Afghanistan, and 6.1 in Iraq.[82] The new US Army's counterinsurgency field manual has followed in the footprints of these density theories, recommending a minimum ratio of 20 troops per 1,000 local residents.[83]

The value of density theory lies in the fact that it illuminates an important aspect of asymmetrical state-nonstate conflicts, serving as a warning that troop density may affect the ability to control and stabilize enemy territory in LICs, and that small but smart militaries may be insufficient for fulfilling that job successfully based merely on their operational and technological sophistication. The theory, however, falls short of constituting a sufficient explanation for success or failure in such conflicts. Each case has its unique circumstances, differing from the other cases in features such as the nature of the societies involved, the spatial extent of the territory, the nature of the terrain, the resources needed for sustaining the conflict, political factors, etc.

The force-to-space ratio relates to the number of troops required to effectively carry out missions within a given territory or to hold a territory.[84] The ratio has affected military operations for decades. It is relatively easy to understand why for an army to be operationally effective in HICs, particularly against more than one enemy or more than one front, a critical mass of troops is crucial. But the recent LICs in Iraq that followed the successful HIC chapter in 2003, or the 2006 Second Lebanon War, have taught us that in asymmetrical conflicts, too, it may be difficult, sometimes even impossible, to destroy a sophisticated guerrilla force by using a limited number of troops, in the hope that RMA warfare would do the job. In LICs a big army may be needed for coping with insurgents trying to compensate for their quantitative inferiority by relying on various force multipliers, such as greater mobility; for defending the border and civilians in frontier areas and in the rear; for capturing the territory from which guerilla warfare is conducted; or for destroying rockets launched by insurgents from within populated areas.

The Complexity of War

The broadening and narrowing of war, which have existed side by side, have not been the only traits of the nature of war. Today's war can also be

---

pubs/2007/RAND_CP22–2003–08.pdf>; James Quinlivan, "Force Requirements in Stability Operations," *Parameters*, Vol. 23 (Winter 1995), pp. 59–69.

82 Ibid.

83 Department of the Army, *FM 3–24 Counterinsurgency*, December 2006, 1–13 <http://www.fas.org/irp/doddir/army/fm3–24.pdf>.

84 Basil H. Liddell Hart, *Thoughts on War* (London: Faber and Faber,1943), p. 62.

characterized by greater complexity. Examples include the notion of "Hybrid Wars," which reflects the obliteration of the borderline between HICs and LICs; Steven Metz's "Gray Area War," which refers to a situation that "involves an enemy [...] that seeks primarily profit, but which has political overtones and a substantially greater capability for strategic planning and the conduct of armed conflict than traditional criminal groups;"[85] the blurred borderlines between military and police operations, one of its typical expressions has been the phenomenon of narco-terrorism;[86] or the emergence of what Chris Bellamy named the empty/saturated battlefield, which is emptier of forces, as a result of the aforementioned reasons, but is saturated with fire.[87]

The Pervasiveness and Importance of LICs

Interstate war has become a rare phenomenon. Some 80 percent of the conflicts during the Cold War were LICs, as were 95 percent of the conflicts that took place in the post-Cold War period.[88] LICs are also a highly significant phenomenon. As mentioned above, despite their name, they can have a devastating effect on the peoples involved; they can threaten the stability of states, subsystems, and the international system as a whole; and they might even negatively affect the future of the nation-state.[89]

The importance of LICs notwithstanding, until the post-Cold War era, strategic thinkers had done an unsatisfactory job in crystallizing a theory of LIC. The focus of 19th century military thinkers and practitioners on symmetrical, "regular" wars is understandable given the fact that in their time, wars of that type dominated the field. Later on, post-World War I strategic thinking was so preoccupied with the introduction of the tank and the plane, with *Blitzkrieg* becoming a central form of warfare that it failed to devote sufficient intellectual resources to attrition and asymmetrical conflicts. Only with the post-World War II wars of national liberation and social revolutions did low-intensity,

85 Steven Metz, *Armed Conflict in the 21st Century: The Information Revolution and Post-Modern Warfare* (Carlisle: US Army War College, April 2000), p. 57.

86 Walter Laqueur, *The New Terrorism, Fanaticism and the Arms of Mass Destruction* (New York: Oxford University Press, 1999), pp. 211–16.

87 Chris Bellamy, *The Future of Land Warfare* (New York: St. Martin's Press, 1987), pp. 274–7.

88 Sivard (ed.), *World Military and Social Expenditures*, pp. 29–31; Ibid., 1989 edition, p. 22; Wallensteen and Sollenberg, "The End of International War?"; Klaus J. Gantzel, "War in the Post-World War II World: Some Empirical Trends and a Theoretical Approach", in David Turton (ed.), *War and Ethnicity: Global Connections and Local Violence* (San Marino: University of Rochester Press, pp. 125–38.

89 Avi Kober, "Low-Intensity Conflicts: Why the Gap between Theory and Practise?" *Defense & Security Analysis*, Vol. 18, No. 1 (March 2002), pp. 15–38.

asymmetrical conflicts receive greater attention. But although the role and status of traditional conventional warfare have been eroded, for a long time both thinkers and practitioners continued focusing on such conflicts rather than sub-conventional ones, failing to break free of their tendency to think in terms of symmetrical, "regular" conflicts.[90] This tendency seems to have a few major explanations. First, HICs were perceived as more military in nature.[91] Second, states, and more so militaries, were used to considering symmetrical challenges more threatening than asymmetrical ones.[92] Third, LICs absorb a broad spectrum of conflicts (such as ethnic conflicts, national liberation struggle, revolutionary war), and the delineations between them are often blurred and combinations whereof are commonplace. Fourth, LICs are multi-dimensional dynamic conflicts, which was yet another source of complexity that was difficult to handle. Fifth, HICs were associated with greater buildup budgets.[93] Sixth, militaries tend towards entrenched traditionalism,[94] which made it difficult for them to adapt to a new reality of war. The result was that the French in Indochina, the Americans in Vietnam[95] or the Soviets in Afghanistan[96] fought with irrelevant and unsuitable doctrines.

---

90 Loren B. Thompson, "Low-Intensity Conflict: An Overview", in Loren B. Thompson (ed.), *Low-Intensity Conflict: The Pattern of Warfare in the Modern World* (Lexington: Lexington Books, 1989), pp. 2–6.

91 Theo Farrell, "World Culture and Military Power," *Security Studies*, Vol. 14, No. 3 (July-September 2005), pp. 448–88; Kober, "Low-Intensity Conflicts."

92 Ashton Carter, "Responding to the Threats: Preventive Defense," paper presented at the conference on 'Challenges to Global and Middle East Security,' Jaffee Center for Strategic Studies and Belfer Center for Science and International Affairs, Herzliah, 15–16 June 1998; *America's National Interests*, Washington, DC: Commission on America's National Interests, 1996). This report distinguishes between "vital," "extremely important," "just important," and "less important" national interests.

93 Deborah D. Avant, *Political Institutions and Military Change: Lessons from Peripheral Wars* (Ithaca: Cornell University Press, 1994), pp. 29–36, 117–29.

94 Liddell Hart, *Thoughts on War*, p. 30; Terry Terriff and Theo Farrell, "Military Change in the New Millennium," in Theo Farrell and Terry Terriff (eds), *The Sources of Military Change: Culture, Politics, Technology* (Boulder: Lynne Rienner, 2001), p. 265. R.A. Mason, "Innovation and the Military Mind," <http://www.au.af.mil/au/awc/awcgate/au24-196.htm>.

95 Craig M. Cameron, "The US Military's 'Two-Front War,' 1963–1988," in Farrell and Terriff (eds), *The Sources of Military Change*, pp. 122–5, 130.

96 The Soviets, for example, used to distinguish between global war, on the one hand, and local war, on the other, making no distinction between counter-insurgency operations and large-scale conventional operations typical of wars such as the Arab-Israeli wars. Stephen Blank, "Soviet Forces in Afghanistan: Unlearning the Lessons of Vietnam," in Stephen Blank et al., *Responding to Low-Intensity Conflict Challenges*, Maxwell Air Base,

When military thinkers eventually recognized the importance of LICs – which happened not earlier than the post-Cold War era – not only has the gap between the centrality of LICs and their representation in security and military studies been closed, a wave of theoretical and empirical works took place that dealt with LICs; for example, literature on ethnic conflicts, Martin Van Creveld's "Non-trinitarian War," Mary Kaldor's concept of "New Wars," Frank Hoffman and James Mattis's "Hybrid War," Frank Hoffman's "Complex Irregular Warfare," or Thomas Hammes and William Lind's "Fourth Generation Warfare (4GW)."[97]

Zeitgeist

Zeitgeist is a non-realistic formative factor at the systemic level. Two aspects of the impact of *Zeitgeist* on military thought are worth relating to. First, military thought tends to reflect the nature of war and strategy in a given period or in a specific intellectual climate, which comes at the expense of its external validity. Second, since the interwar period, and much more so in the wake of the Cold War, moral and legal aspects of war have cast their shadow on military thought, to the point of becoming an integral part of it. This applies mainly to Western democracies.

War in a Given Period

Military thinkers have often had difficulty abstracting themselves from the features of a given period. Their works, therefore, have generally reflected war as seen through the eyes of people living in their own time, imparting to their military thinking a contemporary color. This phenomenon has manifested itself in a number of ways, as some of the following examples will show.

Alabama: Air University Press, 1990, pp. 53–176; Lester W. Grau and Michael A. Gress, *The Soviet-Afghan War: How a Superpower Fought and Lost* (Kansas: University of Kansas Press, 2002).

97 Martin Van Creveld, *The Transformation of War*; Mary Kaldor, *New and Old Wars: Organized Violence in a Global Era* (Cambridge: Polity, 1999); Special issue on Fourth Generation warfare, *Contemporary Security Policy*, Vol. 26, No. 2 (August 2005), pp. 185–285; Frank G. Hoffman, "Complex Irregular Warfare," *Foreign Policy Research Institute E-Notes*, 6 January, 2006; Jeffrey B. White, "Some Thoughts on Irregular Warfare," <http://www.cia.gov/csi/studies/96unclass/iregular.htm>; James N. Mattis and Frank Hoffman, "Future Warfare: The Rise in Hybrid Wars," *Proceedings*, Vol. 132, No. 131 (November 2005), pp. 18–19; Frank G. Hoffman, "Preparing for Hybrid Wars," *Marine Corps Gazette*, Vol. 91, No. 3 (March 2007), pp. pp. 57–61; Frank Hoffman, "Lessons From Lebanon: Hezbollah And Hybrid Wars," *The Evening Bulletin*, 5 September 2006.

The military writings of Sébastien de Vauban reflected the fortification wars that reached their climax during the 17th century,[98] and even Maurice de Saxe, although an exponent of 18th-century warfare, continued to put his faith in fortifications, which in his days continued to be a common and (according to him) preferred form of defense compared to trenches.[99]

The views of Frederick the Great and Maurice de Saxe still reflected the localized and limited character of 18th-century warfare, although in their works one can recognize the initial coming to grips with the evolution from mercenary armies to civilian armies, as well as with questions of morale, motivation and professionalism, which followed in the wake of this process. Jacques Guibert dealt in his later works with the transition, towards the end of the 18th century, from dynastic wars to national wars, pointing out the dangers of the emerging popular armies, which he believed would be unwieldy, unprofessional and inefficient.[100]

Clausewitz, who is rightly considered the foremost theoretical forerunner of the issue of normative and instrumental subordination of military operations and the military echelons to political control and the political echelon, was born into a world in which the state already enjoyed recognized status and power, and in which the use of military force was accepted as a legitimate means for the realization of political objectives. But the fact that the high command in the historical examples that were his principal inspiration – Frederick the Great, Napoleon or the Russian Czar Alexander – was concentrated in the hands of one man, who embodied both the political and the military levels, accounted for Clausewitz's rather simplistic approach to the relationship between politics and war. This resulted in a too-harmonious picture that failed to reflect both the considerable complexity characterizing such a relationship and the extent of bureaucratic politics it entailed.

The most apparent shortcoming in Clausewitz's methodology was the imbalance between the historical cases he examined: he assigned paramount importance to events relating to his own time, and the Napoleonic wars were predominant in the formulation of his theory. This caused Michael Handel to comment that although Clausewitz attempted to formulate a more general theory of war, his work was in the first place an effort to comprehend changes

98 Henry Guerlac, "Vauban: The Impact of Science on War," in Paret (ed.), *Makers of Modern Strategy*, pp. 64–90.

99 Marshal Maurice de Saxe, "My Reveries Upon the Art of War," in Thomas R. Phillips (ed.), *Roots of Strategy* (Harrisburg, Pennsylvania: The Military Service Publishing Company, 1940), pp. 249–55.

100 Palmer, "Frederick the Great, Guibert, Bülow...," pp. 111–12.

that had recently occurred in war, and that "Clausewitz could not have written his theories as we know them before the French Revolution."[101]

It should be added that Clausewitz's tendency to underestimate the importance of technology as a dimension of war and strategy was not only due to the technological resemblance of the European armies during his time but also due to the fact that he failed to read the signs of the Industrial Revolution and its implication. This underestimation of a central dimension of war and strategy seems to be one of the reasons that led him to his stipulation that defense was axiomatically always stronger than offense – a stipulation that has not stood the test of time.[102]

Both Jomini and Clausewitz were deeply affected by the drift towards absolute war, a process that had started during the French Revolution and continued during the Napoleonic wars, and their thinking was dominated by the role of the masses in war. The tension between the notions of absolute war and limited war in Clausewitz's theory no doubt resulted, on the one hand, from his observations of the wars of the French Revolution and the Napoleonic wars, which inspired his formulation of the model of absolute war, and on the other hand from his study of past military conflicts, which represented his model of a limited war.[103]

The works of Bülow, Jomini and Mahan reflected, as mentioned earlier, the growing emphasis on the logistical factor in modern war that was fought between large armies, which in everything connected with armament, equipment, food and other supplies, depended upon large-scale support from the civilian rear. Much of their thinking was devoted to the relationship between bases of operations and lines of operations and logistics.[104]

Marxist thinking on war during the second half of the 19th century reflected the socio-economic situation in general, and the growing class differences in particular, within the industrializing European society.

The trauma of World War I induced a new school of thinkers, such as Liddell Hart and Fuller to raise to the top of their priorities the need to avoid a direct military confrontation. The thinking of aerial warfare theorists in the wake of

---

101 Michael I. Handel, "Introduction," in Handel (ed.), *Clausewitz and Modern Strategy*, p. 4.

102 Handel, "Clausewitz in the Age of Technology."

103 On the distinction between absolute war and limited/real war, see Clausewitz, *On War*, pp. 75–89, 579–81.

104 On Bülow, see Azar Gat, *The Origins of Military Thought* (Oxford: Clarendon Press, 1989), pp. 79–94; on Jomini, see Jomini, Article XVI; on Mahan, see Mahan, *The Influence of Sea Power Upon History*, pp. 1–89.

World War I, in particular Douhet, Mitchell and Seversky,[105] was dominated not only by the emergence in war of a third, aerial dimension, but also by the increasing and intensifying involvement of society in war. Taking both factors into account, they concluded that a future war could be won by dealing a mortal blow to the enemy's centers of population and industry from the air. As pointed out earlier in this chapter, as a result of the post-World War I preoccupation with the mechanization of the battlefield and with airpower, there were hardly any military thinkers who dedicated their thinking in the interwar period to LICs. This would change only after World War II. After World War II, the preoccupation with a nuclear world came at the expense of analyzing conventional warfare.

Intellectual Climate

An important formative factor in military thought has been the underlying intellectual climate that has acted as a source of ideas, viewpoints and methods. This section focuses on two such factors: the neo-classicist school, and the impact of the Enlightenment as compared to Romanticism.

One of the important sources of inspiration of modern military thought is the cultural, social and political framework of the classical world, first of all the "popular republics" of Greece and Rome. Like the ancient habit of mind to create a linkage between past and contemporary themes "to provide a frame in which to puzzle over current problems,"[106] many military thinkers from the 15th century onwards belonged to a school that might be called neo-classicist. The neo-classicists created a direct linkage between the classical and the modern world, almost disregarding intervening developments in the Middle Ages. Many eminent thinkers, from Machiavelli onwards, regarded the Roman legion as an example worth emulating. Machiavelli, a Renaissance man and a forerunner of modern military thought, propagated the application of the Roman legion's structure, organization and tactics to the contemporary army of his time.[107] Machiavelli's admiration was shared by Justus Lipsius and Raimondo

105 On the school of Douhet, Mitchell and Seversky, see Giulio Douhet, *The Command of the Air* (London: Faber & Faber, 1943); Alexander P. de Seversky, *Air Power* (London: H. Jenkins, 1952); Edward Warner, "Douhet, Mitchell, Seversky: Theories of Air Warfare," in Edward M. Earle (ed.), *Makers of Modern Strategy* (New Jersey: Princeton University Press, 1943), pp. 485–503; Bernard Brodie, *Strategy in the Missile Age* (Princeton: Princeton University Press, 1959), pp. 71–106; David MacIsaac, "Voices from the Central Blue: The Air Power Theorists," in Paret (ed.), *Makers of Modern Strategy*, pp. 624–647.

106 Lendon, *Soldiers and Ghosts*, p. 11.

107 This is expressed in Niccolo Machiavelli's *The Art of War* (Indianapolis: Bobbs Merrill, 1965). On Machiavelli's theories, see, for instance, Felix Gilbert, "Machiavelli: The Renaissance of the Art of War," in Paret (ed.), *Makers of Modern Strategy*, pp. 11–31.

Montecuccoli, the two foremost military thinkers of the 16th and 17th centuries.[108]

Frederick the Great and de Saxe, in the 18th century, similarly praised Roman military discipline, which, according to Frederick, was worth emulating, and they, too, marveled at the structure, organization and tactics of the Roman legion, which enabled it to confront successfully even numerically-superior forces.[109] 18th-century thinker Guibert was attracted by both the socio-political and military institutions of the classical popular republics, which he believed to have been a fountainhead of power, quite contrary to the nations and regimes of his own time.[110]

Clausewitz's thinking, too, reflects clear classical influences, but from a different aspect. He was affected by Platonic and Aristotelian views concerning the state, according it a status similar to the Greek polis, in the sense of it representing the general will, and to which all fields of activity were subordinate.[111] The three pillars characterizing war, according to Clausewitz, were, as already mentioned, the people, the military and the government. The people represented the passionate element of war, which was the reason for its being violent and full of enmity and hatred; the military reflected the uncertainty, chaos and friction characterizing the battlefield, which spurred commanders to reveal their creativity and genius; whereas the government represented rationality, the control of the war, and the subordination of war to politics.[112] This division between the people, the military and the government brings to mind the division in the republic as envisaged by Plato, between the guardians (the rulers), the auxiliaries (who defend the state), and the producers.[113]

Other, more modern, sources of inspiration that left their imprint on modern military thought were the ideas of the European Enlightenment during the 17th and 18th centuries and Romanticism in the 19th century. The ideas of the Enlightenment in many ways represented a natural continuation of the scientism of the Renaissance era. Romanticism, in turn, was a reaction – or antithesis – to the Enlightenment, which began to manifest itself in the 19th century, even though early 19th-century military thinkers often continued to

108 Rothenberg, "Maurice de Saxe, Gustavus Adolphus, Rainondo Montecuccoli...," pp. 34–5, 60, 62–3.

109 Palmer, "Frederick the Great, Guibert, Bülow...;" de Saxe, "My Reveries Upon the Art of War."

110 Palmer, "Frederick the Great, Guibert, Bülow...," pp. 107–8.

111 Gat, *Policy and War*, pp. 25–26.

112 Clausewitz, *On War*, p. 89.

113 Plato, *The Republic* (London: Methuen & Co., 1906), in particular Books III and IV; Yehoshafat Harkabi, *War and Strategy* (Tel Aviv: Maarachot, 1990) [Hebrew], p. 388.

be influenced by Enlightenment ideas. As far as military thought is concerned, the Enlightenment provided military thought with a scientific and rationalistic patina, and manifested itself in the desire to formulate principles of war, which would encompass military wisdom within the confines of a few general rules, at the risk of reducing the principles of war to a "manual" for the commander, similar to the laws in the natural sciences.

The most notable examples of this school are Montecuccoli; Vauban, who was referred to earlier in connection with fortification warfare; Guibert; the Prussians Bülow, who raised the use of geometric principles to the level of the absurd; and Frederick the Great.[114]

Military thought between the Enlightenment and Romanticism reflected a blend of the two periods. For example, Jomini stood closer to the Enlightenment than Clausewitz. Though admitting to the artistic nature of warfare, he believed that within the framework of this art it was possible to formulate a set of rules that contained a prescription for the successful conduct of war. Unlike Jomini, Clausewitz rejected this prescriptive and doctrinaire approach to war. He believed that war refuses to be subordinated to a set of rules, and that it has a spiritual, emotional and intuitive side that requires the genius of the commander as a precondition for facing the uncertainties of war.

Morality, Law and War

Another factor that has gained tremendous impact on military thought has been worldwide moral and legal norms. According to the realist tradition, which dominated 19th century military thought, and whose best representative was Clausewitz, morality has almost no relevance to war, and peace means no more than the absence of violence rather than any ideal relationship between two parties. During the interwar period and post-World War II periods, on the other hand, military thought was dominated by Western, liberal thinkers, who preoccupied themselves with the tension between morality and operational effectiveness; more specifically, with questions pertaining to just cause, discriminate use of force, proportionality, civil liberties, and the need to make sure that the use of force enhanced the chances of peace.

114 For an insightful analysis of the influence of the Enlightenment on military thought, see Gat, *The Origins of Military Thought*, pp. 25–94.

The Unit (State) Level

The State's Specific Strategic Conditions

Military thinking tends to reflect the specific strategic circumstances of the thinkers' countries. This is basically a realist, though not a systemic, factor. The association with specific contexts, rather than universal characteristics, causes much military thought to take on the character of a doctrine rather than theory.

Military thought of states with narrow security margins, and/or with inferiority in the force ratios would focus on force multipliers that are expected to compensate for their weaknesses. For example, Prussian military thinkers during the 18th and 19th centuries – Frederick the Great, Moltke and Schlieffen – were first and foremost interested in finding solutions for their country that was surrounded by enemies on several fronts, which could entail a quantitative inferiority vis-à-vis a coalition of enemies. These solutions included, *inter alia*, fighting on interior lines at the strategic level, but on exterior lines at the operational level; offensive approach; indirect approach, first strike, *Blitzkrieg*, and a quick, sequential battlefield decision (on one front after the other).

Douhet did not conceal the fact that the formulation of his ideas with regard to defeating the enemy through aerial bombing of the civilian population and the industrial infrastructure was influenced by the strategic position of Italy, his home country. Italy's enemies were placed along its borders, with a mountain range restricting the mobility of attacking forces on the ground. He conceded that had he been dealing with a conflict between the US and Japan, two countries situated far from each other, his conclusions would almost certainly have been different.[115]

The differences of emphasis between the guerrilla warfare thinkers, too, show how intrinsically related to place was their thinking. Marx and Engels thought in terms of revolutions in industrialized countries such as Germany or England, due to which they assigned a central role to the proletariat in their revolutionary doctrines. Lenin and Mao, on the other hand, due to the very different socio-economic conditions in Russia and China, were constrained to integrate the peasants in the revolutionary process (in the Russian case) or base it on the peasantry (in China).[116]

Yet another example: Anyone familiar with Liddell Hart's writings cannot fail to conclude that his natural inclination would have been in favor of

115 Douhet, *The Command of the Air*, pp. 252–3.

116 See Yehoshafat Harkaby, "From Guerilla Warfare to Guerilla War," in Yehoshafat Harkaby (ed.), *On Guerilla* (Tel Aviv: Maarachot, 1971) [Hebrew], pp. 32–9.

offense.[117] It is therefore difficult to avoid the impression that his opting for defense at the strategic level as the preferred form was, among other things, due to the fact that Britain is an island, and as such it enjoyed a great advantage over its potential invaders, who would have to cross a significant water obstacle first.[118]

Clausewitz represents a somewhat different case. On the one hand, it is true that one of the limitations of his theory lies in the fact that as a Prussian and a continental European he concentrated his research on land warfare, while neglecting naval warfare. But on the other hand, his formative experiences were not confined to the land of his birth, Prussia: he spent extended periods in Russia, as a staff officer in the Czarist army, as well as in French military captivity. This gave him opportunities to broaden his views and observe social and military phenomena as well as strategic questions in a more generalized and open-minded, and far less ethnocentric, manner.[119] Thus, despite his Prussian origins, he did not see the world merely through Prussian-tinted glasses. In this respect Jomini, his contemporary military thinker, equaled Clausewitz, and possibly exceeded him. Jomini, although of Swiss origin, spent most of his life as a staff officer in the French and Russian armies, and thus remained in many ways detached from specific patterns of thought and strategic circumstances associated with any specific country.

Organizational Factors

Promotion Aspirations as a Barrier Against Intellectualism

Norman Dixon argued that a tension exists between promotion aspirations and intellectualism. Commanders tend to believe that in order to qualify for climbing the promotion ladder they had better be void of intellectual skills,

117 Liddell Hart, *Thoughts on War*, p. 291.

118 See, for instance, Donald C. Watt, *Too Serious a Business* (New York: W.W. Norton, 1975), pp. 100–2; John J. Mearsheimer, *Liddell Hart and the Weight of History* (Ithaca: Cornell University Press, 1988), p. 221.

119 On ethnocentrism as a tendency to put one's own country at the center of one's strategic thinking, see: Ken Booth, *Strategy and Ethnocentrism* (New York: Holmes & Meier, 1979); Colin S. Gray, *War, Peace and Victory* (New York: Simon & Schuster, 1990), pp. 44–5; Carl G. Jacobsen (ed.), *Strategic Power: USA/USSR* (New York: St. Martin's Press, 1990), Part 1: "Strategic Culture in Theory and Practice;" Yitzhak Klein, "A Theory of Strategic Culture," *Comparative Strategy*, Vol. 10, No. 1 (January-March 1991), pp. 3-23. Michael Vlahos, "Culture and Foreign Policy," *Foreign Policy*, No. 82 (Spring 1991), pp. 59–78.

and many years spent gaining promotion accentuate characteristics necessary for bureaucratic harmony but alien to intellectual activities.[120]

The Military Mind

Another organizational factor that affects military thought is the so-called military mind. Military thought has often suffered from the tendency of the military mind towards conservatism and entrenched traditionalism on the part of military establishments. For example, Tuchachevsky's vision of deep penetration operations carried out by an army operating as combined-arms teams with motorized infantry, tanks, self-propelled artillery and paratroopers, supported by an air force trained in ground attacks and deep interdiction, was challenged by Defense Commissar Voroshilov and others, who preached defensive operations and more linear tactics, with tanks supporting the infantry.[121] And Douhet's and Mitchell's ideas were met with skepticism and even hostility by the militaries of their respective countries Italy and the US.[122]

"Doers" vs. Thinkers"

Steven Knott criticized the bias among militaries against intellectuals ("thinkers") in favor of individuals of action ("doers"), which is based on the commonly held opinion that intellectuals provide little of practical value and fail in functioning effectively as combat leaders. Such an approach ignores the role played by intellectual talent in the battlefield performance of great commanders such as Joshua L. Chamberlain or George S. Patton. Knott cited Dwight D. Eisenhower's characterization of an intellectual as one "who takes more words than are necessary to tell more than he knows," and cautioned against the marginalization of individual thinkers, which deprives the military of "precious intellectual capital" and of the innovative capacity required to adapt successfully to the changing nature of war.[123]

---

120 For example, Edward N. Luttwak, "On the Need to Reform American Strategy," in Philip S. Kronenberg (ed.), *Planning US Security: Defense Policy in the Eighties* (New York: Pergamon, 1982), p. 23.

121 Condoleezza Rice, "The Making of Soviet Strategy," in Paret (ed.), *Makers of Modern Strategy*, pp. 666–8; John Erickson, *The Soviet High Command* (London: St. Martin's Press, 1962), p. 381; *A Dictionary of Military History* (Oxford: Blackwell, 1994), p. 756 <http://www.marxists.org/history/ussr/government/red-army/1937/wollenberg-red-army/ch07.htm>.

122 Knott, *Knowledge Must Become Capability.*

123 Ibid.

Institutionalized Military Intellectualism

In principle, military thought feeds best on an environment and atmosphere that encourage the preoccupation with the intellectual aspects of the military profession. According to Steven Knot, only institutionalized military intellectualism can achieve successful transformation or, on rare occasion, revolutionize warfare; conversely, individual intellectualism that remains outside of an institutional context is largely impotent.[124] The Prussian/German army nurtured intellectual preoccupation of its commanders since the early 19th century.[125] The Soviet military invested in military thinking both during the interwar period and the Cold War, being affected by a combination of siege mentality, a strong sense of technological inferiority, and philosophical orientation.[126] A report by the US Director of the House Armed Services Subcommittee on Investigations and Oversight praised the Marine Corps for having adopted intellectual standards in the 1980s, which "changed the institution for a decade." The report recommended that Army senior military leaders, too, create "an environment where 'thinking warriors' and 'soldier-scholars' are showcased as the combat leader standard of excellence. [...] Military commanders [...] must [...] reinforce the importance of study and schooling to achieve mastery of the profession of arms. [...] Units should [...] institute officer professional development programs directed toward reading and thinking about military theory and history."[127]

Lesson Learning Processes

Lesson learning processes have also left their stamp on military thought. States and militaries have learned either from the experience of others,[128] or from wars their countries have been involved in. Three formative events during the past two centuries were the Napoleonic Wars and the First and Second World Wars. Lessons drawn from these wars profoundly affected modern military thinking, as several notable examples will show. General Scharnhorst drew several lessons from the Napoleonic wars about the command and control of mass armies, which resulted in the reforms he introduced and led in the Prussian army, centered on its conversion into a conscripted citizen army and

---

124 Ibid.

125 Herbert Rosinski, *The German Army* (New York: Praeger, 1966).

126 Adamsky, "The Conceptual Co-Influence: the Soviet Military-Technical Revolution and the Western Military Innovations."

127 Bamo, "Military Intellectualism."

128 Basil H. Liddell Hart, *Why Don't We Learn from History?* (London: Allen & Unwin, 1972), Part I.

the establishment of a modern General Staff.[129] The lessons of this war, however, became relevant to an entire generation of thinkers and practitioners beyond Prussia, and reforms were also implemented elsewhere. The crucial effect of the lessons from the Napoleonic wars was not only discernible in the works of Jomini and Clausewitz, but influenced officers, who served during the years following the Napoleonic wars and internalized their lessons.

A research conducted by the British and French General Staffs following World War I yielded a number of conclusions about the strength of defense relative to offense. These conclusions received a theoretical value due to the fact that they corresponded with Clausewitz's axiomatic stipulation about defense always being stronger than defense.[130] Additional conclusions were drawn regarding the introduction of mechanized warfare, which influenced British, German, Russian and French thinkers in the interwar period, and induced the German concept of *Blitzkrieg*. Following the introduction of nuclear weapons at the end of World War II conclusions were drawn (which as early as the Korean War were proven to be false) about the irrationality of using military force for achieving political goals in the case of future conflicts.

Naturally, as always is the case when people learn from experience, the possibility arises of a faulty understanding or varying interpretation of the developments concerned. For instance, prior to the outbreak of World War I the General Staffs of the European powers failed to appreciate the technological developments from the mid-19th century onwards, which brought about the ascendancy of defense on the battlefield at the expense of offense. The American Civil War, the Russo-Turkish War, the Boer War, as well as the Russo-Japanese War, all proved insufficient to demonstrate that henceforth maneuver would weaken and gradually lose its dominance in favor of defense, and that future wars would be characterized by attrition and heavy losses in human life and materiel, and would entail a large-scale logistical effort at the military and the national levels.[131]

The Level of the Individual

Individual thinkers have differed with regard to the importance of the intellectual aspects of military thought. As mentioned before, against the backdrop

129 Goerlitz, *History of the German General Staff*, pp. 15–49.

130 See, for instance, Watt, *Too Serious a Business*, p. 69.

131 See Michael Howard, "Men against Fire: The Doctrine of the Offensive in 1914," in Paret (ed.), *Makers of Modern Strategy*, pp. 510–26.

of entrenched traditionalism there have always been thinkers whose curiosity, commitment to innovation, courage to fight for their ideas, and determination have challenged entrenched traditionalism. This section deals with the attitude of different types of individual thinkers towards theory, with spiritual fathers as sources of inspiration for military thinkers, and with the impact of their personal experience on individuals' thought.

Thinkers' Operational Code:[132] Fatalists, "Military Intellectuals," "Intellectual Soldiers," And "Practical Soldiers"

Thinkers have differed with regard to how direct the influence of theory on practice could possibly be, if at all. Most alien to theory are those holding a deterministic/fatalistic view. This view's best manifestation is Tolstoy's *War and Peace*,[133] in which he more than implied that studying war and/or crystallizing doctrines or plans is futile, as the outcomes of military confrontations are ordained by forces beyond human beings' control.

Next come two skeptic approaches. The first one doubts the ability of intellectuals to offer more sensible and better answers to practical challenges than non-intellectuals.[134] The holders of this view contend that a modern army does not need to be wrapped up in theory but rather stick to the nitty gritty of fighting wars. Reflective of such attitude is the so-called practical soldiers' approach. Practical soldiers is a term coined by Liddell Hart. It has a negative connotation, referring to commanders who put their faith in experience and experience-based intuition rather than any intellectually acquired knowledge.[135] Adherents of such an approach believe that in war, like in love, one cannot understand what it is all about until one experiences war personally.[136] A post-modern version of this approach claims that there is no such thing as universal military theory, but rather concrete, specific, context-dependent knowledge.[137]

The second skeptic approach is held by historians, who would prefer to see military organizations "inculcate in their members a relentless empiricism, [and] a disdain for a priori theorizing, if they are to succeed, [... and] cultivate the temperament of the historian [...] rather than the theoretical bent of the

---

132 Stephen G. Walker, "The Evolution of Operational Code Analysis," *Political Psychology*, Vol. 11, No. 2 (June 1990), pp. 403–18.

133 Leo Tolstoy. *War and Peace* (New York: Vintage, 2008).

134 Paul Johnson, *Intellectuals: From Marx and Tolstoy to Sartre and Chomsky* (London: Weidenfeld and Nicolson, 1988).

135 Liddell Hart, *Thoughts on War*, pp. 96–7.

136 Gershon HaCohen, "Educating Senior Officers," in *Is the IDF Prepared for Tomorrow's Challenges?* (Ramat Gan: BESA Colloquia on Strategy and Diplomacy 24, July 2008), p. 33–4.

137 Ibid.

social scientist or philosopher." In other words, it is military history, rather than theory, that indicates the intellectual focus of an army.[138]

Most emphatic to intellectualism in the military field are "intellectual soldiers" and "military intellectuals." "Intellectual soldiers" – to use Morris Janowitz's classification – are officers in active duty, whose primary commitment as commanders is to practice. They are nevertheless inspired by theory and believe in theory-based doctrines and planning. As Janowitz put it, "the intellectual officer [...] brings an intellectual dimension to his job. His intellectual quality is held in check by the needs of the profession. He sees himself primarily as a soldier, and his intellectuality is part of his belief that he is a whole man."[139] According to Steven Knott, while it is not necessary that commanders would be gifted intellectuals, exceptional combat leaders can employ their intellect in solving battlefield challenges.[140]

The "military intellectual" – another Janowitz's category – is a markedly different type. "His attachments and identifications are primarily with intellectuals and with intellectual activities. He would have no trouble shifting from military to university life, for his orientations are essentially scholarly. He is generally denied, or unequipped, for the highest command posts, as would be the case with intellectuals in civilian society. His position is essentially advisory."[141] The military intellectuals' nexus with military thought is affected either by their affiliation with the military as staff officers or as non-uniformed experts who work either with, for or alongside the military, or outside the military, as academics. Military intellectuals' main interest lies with the intellectual aspects of the military profession. One could usually find them contributing to military thought by engaging in producing theory or theory-based doctrines, or by educating both "intellectual soldiers" and "practical soldiers."

The literature offers some similar concepts to those referred to above, for example, "warriors-scholars" and "reflective practitioners." "Warriors-scholars" are grouped into two categories: those who are more "scholars" than "warriors," whose approach to theory is similar to the one held by military intellectuals, and those who are more "warriors" than "scholar," a category that is closer to "intellectual soldiers."[142] "Reflective practitioners" – like "practical soldiers" –

138 Eliot A. Cohen and John Gooch, *Military Misfortunes: The Anatomy of Failure in War* (New York: The Free Press, 1990), 236–7.

139 Janowitz, *The Professional Soldier*, p. 431.

140 Knott, *Knowledge Must Become Capability*, p. 1.

141 Ibid.

142 Andrew Mumford and Bruno Reis, "Constructing and Deconstructing Warrior-Scholars," in Andrew Mumford and Bruno Reis (eds), *The Theory and Practice of Counter-Insurgency: Warrior-Scholars in Irregular War* (New York: Routledge, 2013), pp. 4–17.

learn from their own professional experiences rather than from formal teaching or knowledge.[143]

Most theory adherents have belonged to the principles-of-war school, which expects theory to produce a set of rules of a universal nature, the adoption of which is expected to significantly enhance the chances of success on the battlefield. Among them one can find ancient theorist Sun Tzu; those who were inspired by the 17th and 18th century Enlightenment, like Henry the Duke of Rohan; the Marquise de Silva; Henry Lloyd; and Jomini. 20th century thinkers belonging to the principles-of-war school include Liddell Hart and Fuller.[144]

Unlike the principles-of-war school, Clausewitz represents a more realistic and less pretentious view. He believed that universal prescriptions would be of no avail in light of the unique nature of each war or battle. He therefore perceived of theory as providing the so-called "whole," that is, the systemic order which explains how each part of war relates to another,[145] and as offering explanations for central phenomena in the field of war, so as to influence practice indirectly, by lighting the practitioner's way, training his judgment, and helping him avoid pitfalls.[146] Julian Corbett followed in Clausewitz's footprints, recognizing the principles-of-war's approach limitations. "Strategic analysis can never give exact results. It aims only at approximations, at groupings which will serve to guide but will always leave much to the judgment."[147]

Spiritual Fathers

Many military thinkers have drawn their inspiration from earlier thinkers, not always in a systematic manner. For example, the spiritual father of the earlier mentioned Dutchman Justus Lipsius was Machiavelli; Jomini's ideas were inspired by those of the Welshman Henry Lloyd;[148] Moltke and Schlieffen, as well as Marxist military thinkers, such as Engels or Mao, drew much inspira-

143 Donald A, Schön *The Reflective Practitioner: How Professionals Think in Action* (London: Temple Smith, 1983).

144 For the development of the principles of war, see, for example, John I. Alger, *The Quest for Victory: The History of the Principles of War* (Westport: Greenwood Press, 1982).

145 Clausewitz, *On War*, pp. 183, 484, 577–8. For an excellent discussion of Clausewitz's theory as a *Geschtalt* (a systemic theory), see Handel, *Masters of War*, pp. 345–51.

146 Clausewitz, *On War*, p. 141. See also: Carl von Clausewitz, *The Principles of War* (Harrisburg: The Military service Publishing Company, 1952), p. 11.

147 Julian S. Corbett, *Some Principles of Maritime Strategy* (Annapolis: Naval Institute Press, 1988), pp. 83–4.

148 On Lloyd, and his influence on Jomini, see John Shy, "Jomini,", in Paret (ed.), *Makers of Modern Strategy*, pp. 148–52.

tion from Clausewitz; and Mahan adopted many ideas regarding naval warfare from Jomini's theories on land warfare.

Above all these, Clausewitz stands out for the profusion of non-military thinkers who inspired him, including Plato, Aristotle, Montesquieu, Rousseau, Kant, and perhaps Hegel.[149] Clausewitz had a wide range of interests, not all of which were connected with war and strategy, including art, literature, history and philosophy. These no doubt broadened his horizons and, besides sharpening his intellectual and analytical powers, deepened his insight into everything connected with social phenomena.

Personal Experience

Last but not least, personal experience of thinkers is yet another formative factor at the level of the individual. To mention only a few examples among many: Clausewitz and Jomini were personally involved in the Napoleonic Wars. Clausewitz served as junior as well as senior officer under the colors of both the Prussian and Russian armies, and Jomini served in the French and the Russian Army. The impact of this war on their thinking was mentioned earlier in this chapter. Colonel Ardant du Picq,[150] a French officer during the middle of the 19th century, took part in the Crimean War, the Syrian campaign and two campaigns in Algeria. In all these wars or campaigns he was able to observe the decisive influence of the qualitative factor in war that enabled small European units to defeat numerically larger non European armies. This qualitative aspect formed the crux of his thinking. Engels developed his views on barricade warfare from his personal experience during the German uprisings in 1848–9. [151] Lawrence of Arabia derived many of his theoretical conclusions about guerilla warfare from his experiences during the Arab revolt in the years 1916–18.[152] Jomini and Moltke, each in his own time, fashioned their views about popular warfare, based on their own experience. Jomini, as Marshal Ney's Chief of Staff in Spain, witnessed the disastrous consequences for regular armies of warfare of this kind. By contrast, Moltke, who had observed a similar phenomenon during the Prusso-French War, tried to distinguish between his morally positive approach toward a nation attempting to avoid its

---

149 Gat, *Policy and War*, pp. 25–34. His intellectual development may be traced through Paret, *Clausewitz and the State*, and Aron, *Clausewitz*.

150 Ardant du Picq, *Battle Studies* (Harrisburg, Pennsylvania: The Military Service Publishing Company, 1946).

151 See his introduction in Karl Marx and Friedrich Engels. *Die Klassenkampfe in Frankreich, 1848–1850* (Berlin: Vorwärts, 1895).

152 T.E. Lawrence, *Revolt in the Desert* (New York: G.H. Doran, 1927).

subjection, on the one hand, and the practical conclusion he reached, on the basis of his personal experience, on the other, to the effect that popular uprising has no real chance when confronted by a well-trained and disciplined regular army.[153]

Conclusion

Military thought is affected by both realist and non-realist factors, each offering competing explanations for various aspects of it. For example, a realist explanation for *the cult of technology* attributes it to the availability of sophisticated technologies, whereas a non-realist explanation points to the role played by strategic and military culture in shaping it. In recent years a growing openness can be detected on the part of researchers for combining realist and non-realist factors in models that explain behavior, an approach that may well apply to military thought. It is true, though, that realists insist on treating the realist factors as independent variables, consenting to see the non-realist variables as intervening variables, whereas culturalists might agree to include realist factors as long as they are treated as no more than intervening variables.

Culture is not necessarily confined to individual cases, but can rather be shared by a number of players, as reflected by the pre-World War I cult of the offensive. Moreover, there may exist universal cultures, such as military organizations' tendency towards entrenched traditionalism, a phenomenon that is detrimental to innovation in military thought and practice.

Two systemic factors affect military thought. First, the changing nature of war; second, the so-called *Zeitgeist*. As far as the first factor is concerned, technology has cast its long shadow on almost every aspect of war and strategy. Manifestations of its dominant role and impact on military thought can be easily identified: the need to square the Clausewitzian triangle so as to include the non-material aspects of war; technology becoming an important dimension in Michael Howard's "forgotten dimensions of strategy;" technology being the major reason for the emergence of the operational level of war; the deepening of the battlefield and its expansion to the civilian rear; or the appearance of additional spaces – first the aerial dimension and later on the outer space and the cyberspace. Technology has also always been involved in all kind of revolutions in war in modern time.

153 Jomini, "Summary of the Art of War," p. 445; Jehuda L. Wallach, *Kriegstheorien* (Frankfurt am Main: Bernard & Graefe Verlag Fur Wehrwesen, 1972), pp. 25, 83–4.

Another realist factor at the systemic level has been the broadening of war. This phenomenon has manifested itself in various forms, one of them being the process in which war has receded from the direct battlefield and has involved entire societies, which has caused non-military dimensions of war and strategy to play a major role in military thought and practice. At the same time, however, the focus on the non-operational aspects of war and strategy, like society and technology, has accounted for the neglect of the core aspect of military thought and practice, i.e., the military aspect, as one can learn from the gradual shift of focus from "military studies" before World War II to "strategic studies" and then to "security studies" after World War II.

The combination of the broadening of war and technological developments has created the need to add two levels to the levels-of-war pyramid – the operational level and the grand-strategic level. At the same time it has also been understood that successes on the battlefield were now preferably achieved in a top-down process, wherein the upper levels are expected to create favorable conditions for more efficient actions at the lower ones.

The broadening process has been accompanied by a parallel process of narrowing. Casualty rates during military confrontations have decreased; the size of armies has decreased, thanks to dramatic improvements in weapons performance, their rising cost, better performance and effectiveness, and the need to decrease attrition rates on the battlefield as a result of the ascendancy of firepower; and the control exercised by the highest political and military echelons on operations on the battlefield has tightened due to greater sensitivity to casualties, legal and moral constraints, etc. One of the major repercussions of this process has been the tendency, particularly among highly developed Western countries, to opt for small but smart militaries, and RMA ideas. The narrowing process, however, has been mitigated by the understanding that in order to achieve battlefield success one still needs boots on the ground, and by the fact that in LICs big armies are still needed.

A third process, which has taken place alongside the simultaneous broadening and narrowing of war has been a greater complexity of war, caused by factors such as the obliteration of the differences between HICs and LICs, weak and strong, military and non-military challenges and missions. This complexity has been reflected in the emergence of relatively new notions such as Hybrid War, Fourth Generation Warfare, or the empty/saturated battlefield (saturated with fire but empty from formations and troops).

As for the third systemic factor, i.e., the pervasiveness of LICs in the wake of World War II, one can detect a late reaction to this phenomenon on the part of military thinkers, who have been too slow in reorienting their minds to the new reality. Once they did recognize the importance of LICs, they repeated their former mistake though the other way round. A wave of theoretical and empirical works on ethnic conflicts, non-trinitarian wars (Van Creveld), state-to-nation balance (Benjamin Miller), etc. has swept the security studies discipline during the two recent decades, to the point of overshadowing the preoccupation with HICs. A growing tendency towards post-modern notions with a bent towards LICs has taken place, too, which is reflected in works on Post-Heroic Warfare, Hybrid War, Fourth Generation Warfare, Gray Area War, etc., and the emergence of notions such as "victory image."

Another systemic factor has been *Zeitgeist*. The more *Zeitgeist* has reflected the nature of war in a given period or a specific intellectual climate in a given period, the weaker military thought's external validity has become. Another impact of *Zeitgeist* in recent years has been the unprecedented importance of moral and legal aspects, with notions such as just war, discriminate use of force, proportionality, and civil liberties permeating military thought, particularly of Western democracies engaged in non-existential LICs.

At the unit (state) level, the association of military thought with players' specific strategic conditions rather than universal conditions has caused military thought to focus on doctrine rather than theory. Sometimes a doctrine for a specific player has gained the status of theory, as happened to Douhet' thought on airpower. Other formative factors at the unit level have been organizational factors, such as the tension between promotion aspirations and intellectualism, the tension between doers and thinkers, or the tendency of the so-called military mind towards conservatism and entrenched traditionalism; and the impact of lesson learning from past wars.

At the level of the individual military thought has been affected in various ways:

Fatalists, "military intellectuals," "intellectual soldiers," and "practical soldiers" have held different attitudes towards the importance of military theory; thinkers have often been inspired by spiritual fathers; the personal experience of thinkers has played an important role in their ideas; and a considerable number of thinkers have gained influence thanks to their commitment to innovation and courage in the face of entrenched traditionalism on the part of military organizations in their countries or their fellow military men.

CHAPTER 2

Israeli Intellectual and Modern Focus

In this chapter it is argued, first, that throughout the years the IDF has suffered from lack of intellectualism, which has sometimes had a detrimental effect on its performance. From the 1990s to 2006 the IDF emulated an RMA-inspired American doctrine, which has come at the expense of its originality and innovation, and Israeli military thinking has been affected by false intellectualism and intellectual pretense. These negative trends, however, have been balanced, at least to some extent, by a number of positive aspects, such as vibrant military thinking during Israel's formative years; the relative popularity of military history; the existence of great debates on operational and buildup issues; intellectual efforts to understand attrition; and a more critical attitude toward American thought after 2006.

Second, it is argued that as far as the modern focus of Israeli military thought is concerned, the IDF has been late to adapt to LIC challenges; it has had a strong tactical orientation, although since the 1980s/early 1990s it has eventually recognized the importance of the operational and the grand-strategic levels; it has developed a cult of technology; its traditional force multipliers have been eroded; and logistics has been ascribed a different logic. Finally, moral and legal considerations have become an integral part of Israeli military thought, something that has been typical of the Western world in general.

Intellectual Weakness

Symptoms of Poor Intellectualism in the Military

After having attributed Israeli commanders' successes during the War of Independence to their intellectual skills, among other factors,[1] Prime Minister and Defense Minister David Ben-Gurion expressed his concern in the 1950s about the lack of "intellectual openness" among IDF commanders.[2] And indeed, symptoms of anti-intellectualism in the IDF could be detected throughout the years. Personal testimonies by university professors who taught middle and high rank Israeli commanders, such as Martin Van Creveld from the Hebrew University or Elie Barnavi from Tel Aviv University, found a lack of

1 David Ben-Gurion, *Yichud VeYe'ud* (Tel Aviv: Maarachot, 1971) [Hebrew], p. 43.

2 Eliot Cohen, "An Intellectual Challenge," *Haaretz*, 20 September 1998.

 | DOI 10.1163/9789004306868_003

appreciation among IDF commanders toward the abstract aspects of their profession.[3] After having been invited to lecture before the general staff members, Van Creveld was shocked by their lack of knowledge: "I have never met such a bunch of ignorant people in my entire life. In no other state or organization have I seen people who knew so little about their profession and its theory, including the history and doctrine of their own army."[4] Elie Barnavi was surprised to discover that "these people have not been reading anything besides a daily newspaper for years."[5] Yaacov Hisdai, a senior researcher for the Agranat Commission of Inquiry that investigated the failings of the 1973 October War, reached the conclusion that IDF commanders lagged in innovation, abstract thinking, and a sense of criticism.[6] Uri Milstein, one of the IDF's staunchest critics, has for many years argued that the IDF is an anti-intellectual establishment. According to Milstein, it is no more than an unprofessional, armed militia that employs sophisticated weapon systems. Former editor of *Maarachot*, the IDF's professional journal, claimed that "the average Israeli officer does not read foreign professional literature. Apparently he thinks he does not need it. Hence he cannot update himself about military thinking in the wider world, which is published in dozens of professional journals in many languages."[7] A study by Eliot Cohen et al. affirmed these diagnoses.[8] Only a few years ago, did the Winograd Commission's final report on the functioning of the IDF during the Second Lebanon War reaffirm the existence of a "bad anti-intellectual tendency" among IDF's senior commanders.[9]

Against the backdrop of performance problems during the 1973 October War and the 1982 First Lebanon War, a revolutionary curriculum was launched in 1989 at the Command and Staff College, named *Barak* (lightning, in Hebrew). According to General (retired) Doron Rubin, head of the IDF's TOHAD (doctrine and education branch) – the Israeli TRADOC – the *Barak* program was a

---

3 *Haaretz Weekend Supplement*, 12 April 1996; *Haaretz Weekend Supplement*, 28 March 1997; Giora Eilon, "If We Keep that Way, the State of Israel Might Collapse," Interview with Martin Van Creveld, *Al Hasharon*, 8 March 2002.

4 Eilon, "If We Keep that Way, the State of Israel Might Collapse."

5 *Haaretz Weekend Supplement*, 28 March 1997.

6 <http://www.toravoda.org.il/14hisdai.html>.

7 Eviatar Ben-Tzedef, "The Israel Defense Forces, 1996," *Outpost* (September 1996). Available at <http://www.afsi.org/OUTPOST/96SEP/sep4.htm>.

8 Cohen et al., *Knives, Tanks, and Missiles*, pp. 74–6.

9 The Winograd Commission's final report <http://www.vaadatwino.org.il/pdf/לאינטרנט%20פוסי%20מאוחד.pdf>, pp. 323, 397.

response to the sense that the IDF's middle and senior rank commanders were "ignorant." Rubin explained that since 1973 the IDF had owed its successes on the battlefield to junior commanders, whose education was sufficient for tactical challenges only.[10] *Barak* aimed at higher-rank military commanders, integrating for the first time theoretical, historical, and doctrinal knowledge, while offering tools for improving doctrines, plans, and operational performance.[11] Yet, in 1994, the program was closed. Its instructors, who had been carefully selected from within the IDF as well as from the academia, were blamed for "elitism" and released from their jobs one by one.

In the early 1990s, Chief-of-Staff Ehud Barak decided to suspend the publication of all IDF professional journals except for *Maarachot* and the technology- oriented IAF Bulletin. Particularly baffling was his directive to stop publishing *Tziklon* – a journal dedicated to Hebrew translated professional articles, which had been established three years after the 1973 October War. As *Maarachot* was publishing mostly original materials, and given IDF troops' lack of command of English, *Tziklon*'s task was to fill the void by providing commanders with the best of the professional materials published abroad.[12] Special issues of *Tziklon* dealt with a variety of theories and doctrines, covering topics such as terrorism, nuclear weapons, field intelligence, defense, airborne forces, infantry, anti-tank weapons, mountainous-terrain warfare, naval warfare, aerial strategy, and chemical warfare.

From the mid-1970s to the mid-1980s, the IDF publishing house – also called *Maarachot* – published selected books in translation, authored primarily by classical theorists such as Sun Tzu, Frederick the Great, Clausewitz, Liddell Hart, Mahan, and others. These books were distributed to IDF commanders. As there have not been mandatory reading lists in the IDF, their perusal was left to the discretion of each commander. In the late 1980s, the IDF ceased publication and distribution of these books, for unknown, presumably budgetary, reasons.

Maarachot was launched in 1939, shortly after the outbreak of World War II, as the journal of the *Haganah*, the pre-state Jewish community's military organization. Its goal was to broaden the horizons of the Hebrew soldier and

10 Amnon Lord, "From the Chinese Farm to the Ministry of Defense," <http://rotter.net/cgi-bin/forum/dcboard.cgi?az=show_thread&forum=gil&om=5775&omm=243>.

11 Moshe Shamir, "On Changes in the Inter-Arm Command and Control Training," *Maarachot* 396 (September 2004), pp. 20–5.

12 Avi Kober, "Israeli Military Thinking as Reflected in *Maarachot* Articles, 1948–2000," *Armed Forces & Society*, Vol. 30, No. 1 (Fall 2003), pp. 141–60.

TABLE 2.1 *Military theory 1948-2008 in numbers and percentages*

	N	%
Theory	126	4
Doctrine	474	13
Planning	648	18
No reference to the above topics	2337	65
Total	3585	100

commander.[13] After the establishment of an Israeli regular army, the IDF, in 1948, *Maarachot* became its professional journal. On the occasion of the tenth anniversary of *Maarachot*, in 1949, Prime Minister David Ben-Gurion stressed the need for every Israeli commander to "follow, on an ongoing basis, the new developments in the field of military research and the changing problems of war."[14] This requirement notwithstanding, Israeli military publications in recent decades have exposed the absence of reference to the significant body of literature dedicated to state-of-the-art theory models termed Non-Trinitarian War, Fourth Generation Warfare (4GW), Hybrid War, New War, etc.[15] They have also exposed the IDF's tendency to emulate American doctrines instead of developing original ones, and the neglect of modern military literature in favor of false post-modern notions (see below). The *Maarachot* survey, which covers *Maarachot* books and articles in the period 1948–2008 attests to the continuous poor intellectualism in the IDF. Of the articles published in the

13 *Maarachot* 1 (September 1939), p. 1.

14 Cited in *Maarachot* 270–271 (October 1979), p. 16.

15 Martin Van Creveld, *The Transformation of War* (New York: Free Press 1991); Mary Kaldor, *New and Old Wars: Organized Violence in a Global Era* (Cambridge: Polity, 1999); Special issue on Fourth Generation warfare, *Contemporary Security Policy*, Vol. 26, No. 2 (August 2005), pp. 185–285; Frank G. Hoffman, "Complex Irregular Warfare," *Foreign Policy Research Institute E-Notes*, 6 January 2006; Jeffrey B. White, "Some Thoughts on Irregular Warfare," <http://www.cia.gov/csi/studies/96unclass/iregular.htm>; James N. Mattis and Frank Hoffman, "Future Warfare: The Rise in Hybrid Wars," *Proceedings*, Vol. 132, No. 131 (November 2005), pp. 18–19; Frank G. Hoffman, "Preparing for Hybrid Wars," *Marine Corps Gazette*, Vol. 91, No. 3 (March 2007), pp. pp. 57–61; Frank Hoffman, "Lessons From Lebanon: Hezbollah And Hybrid Wars," *The Evening Bulletin*, 5 September 2006 <http://www.theeveningbulletin.com/site/news.cfm?newsid=17152236&BRD=2737&PAG=461&dept_id=574088&rfi=6>.

TABLE 2.2 *Military theory per period (percentages except for N)*

	1948–1956	1957–1967	1968–1973	1974–1977	1978–1982	1983–1987	1988-1994	1995–2000	2001–2004	2005–2008
N	265	563	544	241	389	393	372	290	247	281
Theory	9.5	1	1	0.5	2	0.5	1.5	6	9	12
Doctrine	25	11	6.5	7.5	12	8	11	21	18	18
Planning	61	27	7	8	8	7	6.5	10	34	28
No reference to the above topics	4.5	61	85.5	84	78	84.5	81	63	39	42
Total	100	100	100	100	100	100	100	100	100	100

journal during this period, only 4 percent dealt with theory (see Table 2.1). Between 1956 and 1994, however, the percentages reached their lowest point, not exceeding 1.5 (see Table 2.2). Of the books published during period only 9.5 percent were dedicated to theory (see Table 2.3).

Other findings of the survey serve as additional testimony to the poor intellectualism of the IDF. Before the 1973 October War, authorship had been the domain of regulars, who had written more than reservists and civilians, combined. Yet, the regulars' share of writing waned following a gradual change in the wake of that war, and in the 1980s, despite various efforts by *Maarachot* editors to encourage original writing by Israeli regular officers. Most of the Israeli authors of *Maarachot*'s articles were non-regulars (see Table 2.4).

This is not unique to the IDF, though (see analysis of *Military Review* publications in the periods 1982–1987, 1995–2000, and 2000–2008 in the Appendix). This is compatible with the following assertion by Gregory Foster, who investigated the education of senior military professionals in the US Army: "It is ironic and disappointing that virtually all the reputed 'experts' on strategic and military affairs familiar to the public are civilian academicians, consultants and journalists."[16]

It is impossible to demand intellectualism without an education and training system which is committed to intellectualism. The IDF's education and training system has in recent decades been the object of much criticism. Critics have blamed it not only for the failure to keep abreast of the dynamic changes in the nature of war, but also in acquainting commanders with the theoretical and historical knowledge necessary for effective functioning in a complex reality of war.[17] Over the years, four Israeli universities – the Hebrew University, Tel Aviv University, Bar-Ilan University, and Ben-Gurion University – have offered the IDF academic programs and degrees in various fields, which have mushroomed into an industry for degrees.[18] Some of these academic programs did not even pretend to focus on military and security studies, and those that had,

---

16 Gregory D. Foster, "Research, Writing, and the Mind of the Strategist," *Joint Force Quarterly*, 11 (Spring 1996), p. 115.

17 Eytan Gilboa, "Educating Israeli Officers in the Process of Peacemaking in the Middle East Conflict," *Journal of Peace Research*, Vol. 16, No. 2 (1979), pp. 155–62; Yehoshafat Harkabi, *War and Strategy* (Tel Aviv: Maarachot 1990), pp. 587–8; Michael Handel, *Masters of War* (London: Frank Cass, 2001), pp. 353–60; Dov Tamari, "Is the IDF Capable of Changing in the Wake of the Second Lebanon War?" *Maarachot*, 415 (November 2007), pp. 26–41.

18 Gabriel Ben-Dor, "The Interface between the Military and the Academic World in Israel," paper presented at an International conference on *'The Decline of the Citizen Armies in Democratic States: Processes and Implications,'* Bar-Ilan University, 18–19 June 2008.

TABLE 2.3 *Status per period (percentages except for N)*

	1948–1956	1957–1967	1968–1973	1974–1977	1978–1982	1983–1987	1988–1994	1995–2000	2001–2004	2005–2008
N	265	563	544	241	389	393	372	290	247	281
Regulars	23	39	37	34	37	25	28	27	30	36
Reservists	3	3	4.5	15.5	16	23	17.5	20	26	28
Civilians	5.5	8.5	11	11	15	21	22	23	34	28
Anonym\ Non-Israelis	68.5	49.5	47.5	39.5	32	31	32.5	30	10	8
Total	100	100	100	100	100	100	100	100	100	100

TABLE 2.4 *Military theory (books) in numbers and percentages*

	N	%
Theory	35	9.5
Doctrine	64	17
Planning	224	60.5
No reference to the above topics	47	13
Total	370	100

offered courses tailored to the areas of interest and research of the particular university's faculty, which covered only a limited scope of topics relevant to the military profession. In his 2006 report, the State Comptroller criticized the preference given by most senior commanders to attend academic programs that provided them with managerial skills rather than military skills.[19]

"Absorptive Imitation" of Foreign Military Thinking

Years ago it was claimed that the IDF had adopted the *Wehrmacht*'s mission-command system (*Auftragstaktik*), that much of what Israel had learned about armored warfare was based on Nazi Germany's experience, or that the IAF was in many respects modeled on the Nazi *Luftwaffe*.[20] It seems that being poorly acquainted with other armies' doctrines, Israeli strategies and tactics in the more distant past were rather experience and intuition-based original creations.[21]

The need to act upon updated doctrines, on the one hand, and poor investment in the intellectual aspects of the profession, on the other, have pushed the IDF towards the emulation of ready, off-the-shelf American doctrines,

---

19 *The State Comptroller's Annual Report* 57A, December 2006, p. 52 <http://www.mevaker.gov.il/serve/contentTree.asp?bookid=474&id=57&contentid=&parentcid=undefined&sw=1024&hw=698>. See also <http://themarker.captain.co.il/hasite/pages/ShArt.jhtml?itemNo=877770>; <http://www.haaretz.co.il/hasite/spages/796743.html>.

20 Eric M. Hammel, *Six Days in June: How Israel Won the 1967 Arab-Israeli War* (New York: Simon & Schuster, 1992), p. 107; Van Creveld, *The Sword and the Olive*, pp. 112, 141, 169; Edward Luttwak and Dan Horowitz, *The Israeli Army* (London: Allen Lane, 1975), p. 121.

21 Luttwak and Horowitz claim that the mission-oriented approach was a uniquely Israeli mode of military operations. Ibid., p. 54.

which it has done in an absorptive rather than competitive manner.[22] The intimate strategic partnership with the US has exposed the IDF to American military thinking, particularly to RMA ideas. These ideas were based on the assumption that technology offered new opportunities for destroying the enemy by using standoff precision fire while saving the lives of troops and minimizing enemy civilians' casualties.[23] Reliance on a large military with the traditional role of capturing enemy territory gave way to emphasis on firepower in general and airpower in particular, and the reliance on the cooperative operations of the Special Forces with the air force. The notion of diffused warfare was adopted, based on the assumption, which was shared by many RMA thinkers, that a fundamental shift had taken place from campaigns consisting of horizontal collisions between rival forces, breaking through the opponent's layers of defense and conducted along distinct lines with distinct starting and end points, to a diffused confrontation carried out simultaneously over the entire "battlespace," in which the force's mass is distributed to a multitude of separate pressure points rather than concentrated against a center of gravity.[24]

Constituting a central component of a network-centric joint but "defused" warfare,[25] one of the roles of the Special Forces was to shorten the sensor-to-shooter loop to near-real time and create various kinds of "effects." Not only was the RMA-inspired notion of Effects-Based Operations (EBO) elusive, [26] by

---

22 See for example, Shamir, "When Did a Big Mac Become Better than a Falafel?" On the distinction between imitation and assimilation, see Ahad Haam's "Imitation and Assimilation," in Ahad Ha'am, *Essays, Letters, Memoirs* (Oxford: East and West Library, 1946), pp. 71–5.

23 For Chief-of-Staff Shaul Mofaz's perception of the technological opportunities, see: Stella Koren-Lieber, "A Small but High-Tech Military," *Globes*, 29 August 1999. Mofaz was the Chief-of-Staff under whose command the IDF completed Aviv Ne'urim.

24 Haim Assa and Yedidya Yaari, *Diffused Warfare* (Tel Aviv: Yediot Aharonot, 2005) [Hebrew]. See also Alex Fishman, "They Extinguished the Fire and Gained Time," *Yediot Aharonot Weekend Supplement*, 21 January 2005.

25 Chris C. Demchak, "Complexity and Theory of Networked Militaries," in Theo Farrell and Terry Terriff (eds), *The Sources of Military Change: Culture, Politics, Technology* (Boulder: Lynne Rienner, 2002), pp. 221–62; Lieutenant-Colonel Edmund C. Blash, USAR, "Signal Forum: Network-Centric Warfare Pro & Con," <http://www.iwar.org.uk/rma/resources/ncw/ncw-forum.htm>; Lieutenant- General William S. Wallace, "Network-Enabled Battle Command," <http://www.rusi.org/downloads/pub_rds/Wallace.pdf>.

26 Effect is defined as "the physical, functional, or psychological outcome, event, or consequence that results from specific military or non-military actions. EBO are defined as "a process for obtaining a desired strategic outcome or effect on the enemy through the synergistic and cumulative application of the full range of military and non-military capabilities at all levels of conflict." Lieutenant Colonel Allen W. Batschelet, *Effects-based*

adopting it the IDF senior commanders have distanced themselves from a clear definition of the mission and from the old but simple notion of center of gravity, which united military thinkers and practitioners for centuries. "Swarming," that is, coordinated attacks from different directions by small, dispersed, and networked units – an American concept as well[27] – was also enthusiastically adopted by the IDF.[28]

Another expression of the emulation tendency has been the treatment of the notion of "jointness." American thinking views joint operations as ones carried out by forces consisting of more than one arm/corps/civilian agency headed by a united command, and characterized by mutual relationship between the forces' components. The relative weight of each component is supposed to be dynamic and to vary according to the nature of the particular missions ahead. Jointness is also perceived as a means of curtailing the sectorial tendency of the military. But there is a difference between the relevance of jointness for the US and for Israel, at least with regard to two aspects. First, the US needs jointness especially for carrying out overseas counter-insurgency (COIN) operations, where task forces or expeditionary forces are used, whereas this is not the case as far as the IDF is concerned. Second, for many years now the IDF's ground forces have been the dominant arm, and in recent decade their status as such has been challenged and eroded by the IAF. This is incompatible with the assumption that the relative weight of the forces' components is dynamic.[29]

As was mentioned in Chapter 1, there is no such thing as a universal doctrine, as doctrine, unlike theory, has to suit the specific state's or army's particular strategic conditions, which usually vary from one context to another.

---

operations: A New Operational Model? (Carlisle: US Army War College, 2002), p. 2. See also Edward A. Smith, *Effects Based Operations: Applying Network-centric Warfare in Peace, Crisis, and War* (Washington, DC: Department of Defense Command and Control Research Program, 2002). On its adoption by the IDF, see The Winograd Commission's Interim Report <http://www.vaadatwino.org.il/pdf/לאינטרנט%20סופי%20מאוחד.pdf>, p. 49.

27 Sean J.A. Edwards, *Swarming on the Battlefield: Past, Present, and Future* (Santa Monica: Rand, 2000); John Arquilla and David Ronfeldt, *Swarming & the Future of Conflict* (Santa Monica: Rand, 2000); Amir Rapaport, *Friendly Fire* (Tel Aviv: Maariv, 2007) [Hebrew], p. 81.

28 Gal Hirsch, "On Dinosaurs and Hornets: A Critical View on Operations Moulds in Asymmetric Conflicts," *RUSI Journal*, Vol. 148, No. 4 (August 2003), p. 63; Ofer Shelah and Yoav Limor, *Captives in Lebanon* (Tel Aviv: Yediot Aharonot, 2007) [Hebrew], pp. 197–8.

29 Dov Tamari and Meir Klifi, "The IDF's Operational Conception," *Maarachot* 423 (February 2009), pp. 26–41; Zvi Lanir, "Who Needs the Concept Jointness," *Maarachot* 401 (June 2005), pp. 20–27.

Emulation of military doctrines, however, is a well-known phenomenon. For example, it was claimed that in the late 1970s, the American AirLand battle (ALB) doctrine was nothing more than an emulation of the interwar Soviet deep-battle doctrine, or that RMA thinking had its roots in the Cold War Soviet military technological revolution (MTR).[30] Ironically, in the Israeli case, the IDF joined the emulators' club after having ridiculed the Arab armies' limitations in forming their own original doctrines, ascribing this to a lack of independent or innovative military thinking.

With the passing years more and more IDF officers have attended American military colleges and centers like the Army War College, the US Air Force Academy, or the Army Combined Arms Center, and others have studied at the National Defense University (NDU), the Johns Hopkins' School for Advanced International Studies (SAIS), or the John F. Kennedy School of Government at Harvard.[31] This is yet to be proved, but it may have strengthened the American impact on Israeli military thinking.

False Intellectualism and Intellectual Pretense

During the 1990s false intellectualism started emerging in the IDF. Three examples come to mind. First, in the late 1990s, the IDF's general staff launched a reform to ensure greater efficiency in the organization. In the center of this reform, named *Aviv Neurim* (in Hebrew, spring of youth), stood greater authority for the arms and field units; unification of thc command and budgetary authorities; the creation of a unitary body in charge of building up the ground forces; and a commitment to provide "more security for each Israeli Shekel spent." Not only did the reform lack any ambition to improve military thinking, it was oriented toward producing better managers rather than great captains, reflecting deep misunderstanding of the difference between the skills required from those operating in the non-linear, often paradoxical battlefield, on the one hand, and managerial skills, which are insufficient and may even be detrimental on the battlefield, on the other.[32] *Aviv Neurim* created a false impression

30 Adamsky, "The Conceptual Co-Influence," paper presented at the IAIS annual meeting, Hebrew University, 6 June 2006; Dima Adamsky, *The Culture of Military Innovation: The Impact of Cultural Factors on the Revolution in Military Affairs in Russia, the US, and Israel* (Stanford: Stanford University Press, 2010).

31 Stuart Cohen, "Light and Shadows in US-Israel Military Ties, 1948–2010," *in* Robert Freedman (ed.), *Israel and the United States: Six Decades of US-Israeli Relations* (Boulder: Westview Press, 2012), p. 157.

32 Aharon Zeevi, "Aviv Ne'urim: The Vision and its Implementation," *Maarachot* 358 (April 1998), pp. 3–6.

that the IDF was experiencing an intellectual renaissance. In fact, the reform's spiritual fathers and heroes were none other than organizational advisers, most of whom had no military background or understanding whatsoever.

The second example reflects a similar state of mind. When Shaul Mofaz was chief-of staff in the late-1990s, he distributed Spencer Johnson's *Who Moved My Cheese?* to IDF commanders. The book's natural target audiences were managers rather than military commanders. The intention was supposedly good, namely, helping commanders overcome fears of change and cope with changing realities. Putting aside the debate on this particular book's value, however, it constituted a huge change from the 1970s and the 1980s, when IDF commanders received classical military theory books.

The third example pertains to post-modern ideas that were proliferated among IDF senior commanders by the Operational Theory Research Institute (OTRI), which had been established in 1995. OTRI was supposed to develop operational art knowledge that would be useful for commanders at the operational level,[33] and dubbed by its former chief researcher Shimon Naveh as no less than a "new curing gospel."[34] Alongside the adoption of a relatively solid, though ready, off-the-shelf, Soviet-inspired American operational thinking, under the auspices of that institute many IDF commanders underwent a process of superficial intellectualization, pretending to amend the anti-intellectual tendencies in the military. During the 1990s and early 2000s, OTRI's heads embraced non-military post-modern theories, claiming that this would equip senior officers with the tools necessary for dealing with the complex and changing realities of war. Classic military thinkers seem to have become no more than names whose sayings were occasionally cited but whose writings were not read or studied in depth.[35] According to OTRI's head Naveh,

> We read Christopher Alexander [...], John Forester, and other architects. We are reading Gregory Bateson; we are reading Clifford Geertz. [...] Our generals are reflecting on these kinds of materials. We have established a school and have developed a curriculum that trains 'operational architects.'" In his lectures, Naveh used a diagram resembling a "square of

33 Tamir Libel, "IDF Operational Level Doctrine and Education During the 1990s," *Defense and Security Analysis*, Vol. 26, No. 3 (September 2010), pp. 321–4.

34 Shimon Naveh, *Operational Art and the IDF: A Critical Study of a Command Culture* (Washington, DC: Center for Strategic and Budgetary Assessment, 2007), p. 3.

35 Eyal Weizman, "Israeli Military Using Post-Structuralism as Operational Theory," Infoshop News, 1 August 2006 <http://www.infoshop.org/inews/article.php?story=20060801170800738>. See also Yotam Feldman, "Warhead," interview with Shimon Naveh, *Haaretz Supplement*, 26 October 2007.

> opposition" that plots a set of logical relationships between certain propositions referring to military and guerrilla operations. Spouting phrases such as "difference and repetition – the dialectics of structuring and structure," "formless rival entities," "fractal maneuver," "velocity vs. rhythms," "the Wahhabi war machine," "post-modern anarchists" and "nomadic terrorists," Naveh and his team often evoked the work of Deleuze and Guattari. "War machines," according to these philosophers, "are polymorphous, diffuse organizations characterized by their capacity for metamorphosis, made up of small groups that split up or merge with one another, depending on contingency and circumstances.[36]

In an interview he elaborated on the way post-modern philosophy could be applied to military matters:

> Several of the concepts in French philosopher Gilles Deleuze and psychoanalyst Félix Guattari's *A Thousand Plateaux* became instrumental for us [...] allowing us to explain contemporary situations in a way that we could not have otherwise. It problematized our own paradigms. Most important was the distinction they have pointed out between the concepts of "smooth" and "striated" space [which accordingly reflect] the organizational concepts of the "war machine" and the "state apparatus". In the IDF we now often use the term "to smooth out space" when we want to refer to operation in a space as if it had no borders. [...] Palestinian areas could indeed be thought of as "striated" in the sense that they are enclosed by fences, walls, ditches, roads blocks, and so on.

Naveh interpreted moving through walls, as the IDF did during the battle in Nablus in 2002, as application of this philosophy: "In Nablus the IDF understood urban fighting as a spatial problem. [...] Travelling through walls is a simple mechanical solution that connects theory and practice." Aviv Kokhavi, one of Naveh's disciples, who was commander of the Paratrooper Brigade during this battle, explained the principle that guided him:

> The space that you look at, this room that you look at, is nothing but your interpretation of it. [...] The question is how do you interpret the alley? [...] We interpreted the alley as a place forbidden to walk through and the door as a place forbidden to pass through, and the window as a place forbidden to look through, because a weapon awaits us in the alley, and a

---

36 <http://jdeanicite.typepad.com/i_cite/2006/09/why_the_israeli.html>.

> booby trap awaits us behind the doors. This is because the enemy interprets space in a traditional, classical manner, and I do not want to obey this interpretation and fall into his traps. […] I want to surprise him! This is the essence of war. I need to win […]. This is why that we opted for the methodology of moving through walls. […We acted] like a worm that eats its way forward, emerging at points and then disappearing. […] I said to my troops, "Friends! […] if until now you were used to move along roads and sidewalks, forget it! From now on we all walk through walls![37]

Brigadier-General Gal Hirsch, another disciple, repeated the same rationale when referring to urban warfare during the 2002 Operation Defensive Shield:

> When we launched Operation Defensive Shield, the Paratroopers Brigade operated simultaneously from all directions. To circumvent the traps that Force 17, the Palestinian Authority's (PA) military intelligence, and many other PA forces joined by terrorist organizations operating on the ground had set for us, and to escape similar snares on the streets, we resorted to a unique strategic measure developed specifically for Operation Defensive Shield: the hammer, which we used to tear down walls and be able to move from house to house selectively and judiciously.[38]

Chief-of-Staff Moshe Yaalon was one of the great proponents of OTRI's approach. "[We in the IDF] should preoccupy ourselves with architecture. [We must incorporate design in the pre-planning estimate-of-the-situation process] before we get to action […]. Design is about culture, perception, and reflection. [In order to be relevant,] it requires awareness of the context [we are operating in]."[39]

The main practical damage inflicted by OTRI was its use of flowery "neologisms" by the institute's instructors and graduates, which the Winograd

---

37 Eyal Weizman, Interview with Shimon Naveh, *Frieze*, No. 99 (May 2006) <http://www.frieze.com/issue/article/the_art_of_war/>. See also Weizman, "Israeli Military Using Post-Structuralism as Operational Theory;" Eyal Weizman, "Walking Through Walls," European Institute for Progressive Cultural Policies 01 (2007) <http://eipcp.net/transversal/0507/weizman/en>; Eyal Weizman. "Lethal Theory," Open 2009/No.18/2030: War Zone Amsterdam <http://www.skor.nl/_files/Files/OPEN18_P80-99(1).pdf>.

38 Gal Hirsch, "Urban Warfare," *Military and Strategic Affairs* (April 2014), p. 25 <http://www.inss.org.il/uploadImages/systemFiles/HirschUrbanWarfare.pdf>.

39 Moshe Yaalon, *A Long- Short Way* (Tel Aviv: Yediot Aharonot, 2008) [Hebrew], p. 110

Commission's final report criticized as creating a "tower of Babel" instead of a clear, common language and mutual understanding among commanders, particularly between commanders at the operational and tactical levels.[40]

Since OTRI's dissolution, the IDF has been trying to make the best out of the 2006 Second Lebanon war's failures. A new think tank – the Dado Center (named after the late Chief-of-Staff David (Dado) Elazar – was established in OTRI's place in 2007. The center's mandate has been similar to OTRI's, namely, to develop knowledge and educate commanders in the field of operation art, but it has implicitly been committed to a more solid approach.[41] Under the leadership of Chief-of-Staff Gabi Ashkenazi the IDF was ordered to return to the classical military terms. For example, the aforementioned elusive term of "Effect Based Operation" was replaced by a concrete definition of the "expected achievement" on the battlefield; all arms and corps were ordered to use the same terminology; and ground maneuver was to lead the effort to achieve battlefield success, backed by the other arms.[42]

A Number of Positive Symptoms

Nevertheless, to balance the picture, at least to some extent, it should be noted that military thinking in Israel has not been totally bereft of intellectualism. First, Israel's formative years, both before and after independence, saw a relatively extensive preoccupation with theoretical and doctrinal issues by *Hagahah* and later on IDF commanders. Second, in contrast with the overall little interest in theory, military history has been rather popular with researchers and commanders, particularly prior to the 1973 October War. Third, Israeli military thinking can take credit for a number of great debates that were held over the years on major military issues. The following sections will elaborate on these aspects.

Vibrant Military Thinking during the Pre-State Years

Between the 1920s and the 1940s, a great effort was made by the Jewish security establishment in pre-State Palestine to study foreign armies' strategies and tactics, and to educate commanders in their light. In particular, the *Haganah*

40 The Winograd Commission's final report, pp. 274–5, 318, 321–2.

41 Interview with Dr. Eitan Shamir, formerly researcher at the Dado Center, 23 March 2014.

42 Abe F. Marrero, "The Tactics of Operation Cast Lead," in Scott C. Farquhar (ed.), *Back to Basics: Study of the Second Lebanon War and Operation Cast Lead* (Fort Leavenworth: US Army Combined Arms Center, May 2009), pp.83–102; Amir Rapaport, "Back to the Traditional, Vernacular Language," *NRG*, 12 October 2008 <http://www.nrg.co.il/online/1/ART1/798/019.html>.

learned from the British, German, Austrian and American thought and practice, although in quite an eclectic manner. Some of the individuals who were involved in this process, such as Yehoshua Globerman, the *Haganah*'s Chief of Training, or Yitzhak Dubno, the Chief of Training of the *Palmach*, the *Haganah*'s elite fighting force, lacked any military background, but thanks to their analytic skills they turned to be leading figures with an original, innovative military thinking (see Chapter 5).[43] Between 1922 and 1924, 150 *Haganah* commanders graduated a training program that took place in Kibbutz Ein Harod, where they were exposed to German theoretical and doctrinal thought. Later on, during the 1930s, instructors in officer courses taught the cadets German and British military thought. They became particularly familiar with British sources as many Jews had served in the British army during and after World War I, as Jewish policemen. Others served in the Zion Mule Corps on the Gallipoli front, or took part in Orde Wingate's Night Squads.

Elazar (Lasia) Galilli, editor-in-Chief of *Maarachot* since its establishment in 1939, testified that he himself had taught both *Palmach* and non-*Palmach* senior *Haganah* commanders military theory and doctrine, citing the names of top commanders such as Yaacov Dori (who would later become the IDF's first Chief-of-Staff), Shaul Avigur, Yitzhak Sadeh, Yigal Allon, Shlomo Shamir, and Eliyahu Cohen.[44] But during the 1940s, the *Haganah* was less interested in regular warfare theories and doctrines or military professionalism, focusing instead on irregular warfare tactics, and the spirit of combat, and the *Palmach* preferred "force" to "brain," and was not ashamed of its ignorance, as far as military theory was concerned.[45] Once a regular army was established with the founding of the State of Israel, Prime Minister David Ben Gurion encouraged the integration in the IDF of commanders who had military background of regular warfare after having served in the German, British, Austrian, Russian or American armies. It was therefore only natural for him to offer senior positions to Yohanan Ratner (ex-brigade commander in the Russian army), Fritz Eshet (a German army veteran), David Marcus (ex-deputy division commander in the US army during World War II), Fred Harris (ex-American officer), Mordechai (Monty) Green (former lieutenant-colonel in the British Army), Efraim

43 Nahum Bogner, *Military Thought in the Hagahah* (Tel Aviv: MOD, 1998) [Hebrew], pp. 166–70.

44 Transcript of an interview held by *Maarachot* editors with Lasia Galilli (no date mentioned).

45 Alon Kadish, *La-Meshek Vela-Neshek: Ha-Hakhsharot Ha-Meguyasot Ba-Palmah* (Ramat Ef'al: The Center for the History of the Hagahah, 1995) [Hebrew], pp. 147–8, 150–52, 157–9.

Ben-Artzi (former lieutenant-colonel in the British army in Palestine), or Haim Laskov (major in the British army). Some of these persons served in the IDF as staff officers, but others, like Marcus and Laskov, functioned as combat commanders.

Vibrant Military Thinking during Israel's Early Years and a Renewed Interest in Theory since the Mid-1990s

Israel's early decades, particularly its formative years – 1948–1956 – saw significant preoccupation with theoretical and historical issues. This is reflected in *Maarachot* pieces, 9.5 percent of which were dedicated to theory (see Table 2.2), and 43.5 percent to military history (see Table 2.6). This great interest in military theory and history can be explained by the existence of the aforementioned hard core of foreign-language speaking commanders and staff officers, who were exposed to foreign armies' military thought.

In the interim period between the State's early years and the 1990s, Israeli military thought focused on doctrines rather than theory. Noteworthy within this period is the updating of the IDF's tactical doctrine in the 1980s, and the initial steps taken towards the crystallization of an operational doctrine in the mid-1990s. As already pointed out, the new doctrinal thinking, however, developed into a dangerous blend of an emulated, ready, off-the-shelf, American doctrine and pretentious post-modern, sometimes completely outlandish, ideas.

A renewed interest in theory characterized the 1990s and the 2000s. According to the *Maarachot* survey, 6 percent of *Maarachot* pieces focusing on theory published in the period 1994–2000, 9 percent in 2000–2004, and a 12 percent peak in the period 2004–2008 have constituted remarkable landmarks (see Table 2.2). The explanation for this seems to lie in the changes that had taken place in the nature of war since the end of the Cold War and the internalization by the IDF of the central role played by LICs, after years of professional attention focused on HICs, and the recognition of the need for a deeper understanding of this kind of war.

The Relative Popularity of Military History

If one shares the view by Eliot Cohen et al that it is military history rather than theory that attests to the intellectual focus of an army,[46] then contrary to the assessment of the IDF as a non-intellectual army, the IDF should be held quite as the opposite. 17 percent of the articles published in *Maarachot* from 1948 to 2008 referred to particular events related either to the history of the

46 Cohen et al., *Knives, Tanks, and Missiles*, pp. 74–6.

TABLE 2.5 *Military history 1948–2008 in numbers and percentages*

Military History 1948–2008	N	%
History	594	17
No reference	2991	83
Total	3585	100

TABLE 2.6 *Military history per period (percentages except for N)*

	1948–1956	1957- 1967	1968–1973	1974–1977	1978–1982	1983–1987	1988–1994	1995–2000	2001–2004	2005–2008
N	265	563	544	241	389	393	372	290	247	281
History	43.5	16.5	17	11.5	9.5	13	7	11.5	24	20
No reference	56.5	83.5	83	88.5	90.5	87	93	88.5	76	80
Total	100	100	100	100	100	100	100	100	100	100

Arab-Israeli wars or to military history in general (see Table 2.5). Similar to the interest in theory, during the State's early years (1948–56) 43.5 percent of *Maarachot* articles dealt with military history. Until 1973, the percentages ranged between 16.5 percent and 17 percent. After 1973 and until the 2000s, there was a sharp decline in the interest in military history, but in the 2000s this interest rose once more, above 20 percent, a trend compatible with what we saw with military theory, though in smaller percentages (see Table 2.6).

The Establishment of the Tactical Command College

In 1999, a Tactical Command College was established by the IDF, marking a breakthrough in the professional education and training of Israeli commanders. The idea was to improve the professional education of the land forces' junior officers, based on the assumption that this would make them better commanders and compensate, at least to some extent, for their lack of combat experience. Skeptics, both then and since, on the other hand, have expressed their preference for educating commanders through example set by their superiors, lots of training, and daily experience gained within their units.[47]

Great Debates

Great debates have been held throughout the years on major force buildup, operational and intelligence issues, among them: the debate in the early 1950s on the tank versus the infantry as the dominant element in the ground forces;[48] the debate in the early 1950s on airpower, between those who advocated the destruction of enemy aircraft in the air and those who believed in air strikes against enemy aircraft on the ground as a means for obtaining air superiority;[49] the debate between the protagonists of elastic, mobile defense and those who believed in a combination of static and mobile defense in the Sinai after 1967;[50] the debate on the role played by the tank on the battlefield that was held in the mid-1970s/early 1980s, in the light of the 1973 October War;[51] the debate between the offensive approach, on the one hand, and a more balanced

47 Uzi Ben-Shalom and Gideon Sharav, "The Military Profession in Israel," *Maarachot* 441 (February 2012), pp. 28–36.

48 Luttwak and Horowitz, *The Israeli Army*, pp. 126–32; Haim Laskov and Meir Zorea, "Should One Wage War," *Maariv*, 10 October 1965.

49 Luttwak and Horowitz, *The Israeli Army*, 121.

50 Ibid., p. 318.

51 Yehuda L. Wallach, "Obits for the Tank," *Maarachot Shiryon* 23 (July 1971), pp. 40–1; Moshe Bar-Kochva, *Chariots of Steel* (Tel-Aviv: Maarachot, 1989) [Hebrew], pp. 573–6; Israel Tal, "The Tank at Present and in the Future," *Maarachot* 281 (November 1981), pp. 2–7; Micha Bar, "The Tank's Obscure Future," *Maarachot* 339 (February 1995), pp. 2–9.

approach, on the other, in the early 1980s, as a result of the 1973 October War;[52] competing explanations for the 1973 intelligence failure – from cognitive-psychological explanations to organizational ones;[53] the firepower/maneuver debate as a result of the development of Precision Guided Munitions (PGMs) in the 1980s;[54] the debate over the Lavi aircraft during the 1980s;[55] the debate over active defense against ballistic missiles in the 1990s;[56] the People's army versus an all-volunteer force controversy, since the 1990s;[57] the debate on the criteria for force design, between the traditional counter-threat approach and the relative advantage approach, in the late 1990s;[58] or the more recent debate over the impact of RMA on the IDF.[59]

The difference between the way commanders, on the one hand, and academics or officers with scientific background, on the other, presented their arguments in some of the debates has demonstrated how significant the contribution of theoretical tools could be for systematic analysis of military and security problems. Arguments by Reuven Pedatzur on the Lavi, Isaac Ben-Israel on intelligence estimates, Ben-Israel and Dan Schueftan on force design

52 The staunchest offense advocates were Israel Tal and Dov Tamari. Israel Tal, "Israel's Security Doctrine," *Maarachot* 253 (December 1976), pp. 2–9; "Offense and Defense in the Wars of Israel," *Maarachot* 311 (March 1988), pp. 4–7; Dov Tamari, "Offense or Defense: Do We Have a Choice?" *Maarachot* 289–290 (October 1983), pp. 5–11. For a balanced approach, see Levite, *Offense and Defense in Israeli Military Doctrine.*

53 Uri Bar-Joseph, *The Watchman Fell Asleep* (Tel-Aviv: Zmora-Bitan, 2001) [Hebrew], pp. 22–27.

54 Amnon Yogev, "Israel's Security in the 1990s and Beyond," *Alpayim*, Vol. 1 (June 1989), pp. 166–85.

55 Aaron S. Klieman, "Lavi: The Lion Has Yet to Roar," *Journal of Defense and Diplomacy*, Vol. 4 (August 1986), pp. 22–9; Gerald M. Steinberg, "Lessons of the Lavi," *Midstream*, Vol. 33 (November 1987), pp. 3–6.

56 Arieh Stav and Baruch Koroth (eds), *Ballistic Missiles: The Threat and Response* (Tel-Aviv: Yediot Aharonot ,1999) [Hebrew]; Yiftah S. Shapir, "Non-Conventional Solutions for Non-Conventional Dilemmas?" *Journal of Strategic Studies*, Vol. 24, No. 2 (June 2001), pp. 153–63; Reuven Pedatzur, *The Arrow System* (Tel-Aviv: Jaffee Center for Strategic Studies, 1993).

57 For various opinions on this issue, see Stuart Cohen (ed.), *The New Citizen Armies* (London: Routledge, 2010).

58 Isaac Ben-Israel, "The Military Buildup's Theory of Relativity," *Maarachot* 352–353 (August 1997), pp. 33–42; Dan Schueftan, "Beyond the Relative Advantage," *Maarachot* 356–357 (March 1998), pp. 70–79.

59 Cohen et al., *Knives, Tanks, and Missiles*; Chris C. Demchack, "Technology's Burden, the RMA, and the IDF: Organizing the Hypertext Organization for Future 'Wars of Disruption'?" *Journal of Strategic Studies*, Vol. 24, No, 2 (June 2001), pp. 77–146.

criteria, and Stuart Cohen and Yagil Levy on the obsoleteness of the citizen army model – have illustrated the value of theoretical tools.[60]

Intellectual Efforts to Understand Attrition

As Israel has been engaged in wars of attrition since the early 1950s, one would expect its military thinking to offer theory-based and experience-based doctrines, as well as strategies and tactics to cope with attrition challenges. The emergence of an implicit Israeli attrition conception, however, has taken place very slowly and gradually. It has been based on the following assumptions and principles: As wars of attrition serve the Arabs, while Israel prefers short and decisive wars, Israel will not initiate wars of attrition; since the 1980s asymmetrical wars of attrition have no longer belonged in the "current security" category, as they have been challenging Israel's "basic security;" asymmetrical wars of attrition have become "battles of conviction," affecting the enemy society's consciousness rather than its capabilities; the chances of achieving decisive victory via attrition are fairly slim; the challenge of attrition cannot be dealt with in one *Blitzkrieg*-style attempt, but rather via a cumulative multidimensional process; in asymmetrical contexts in particular, an offensive approach has to be complemented by a defensive approach; and abiding by high moral standards is important for both ethical and pragmatic reasons. Post-heroic conduct of war has become a *modus operandi* in Israel's asymmetrical wars of attrition, bridging operational effectiveness, on the one hand, and casualty aversion and morality, on the other (this aspect is elaborated on later in this chapter).[61]

Against the backdrop of the Intifadas, and to a great extent triggered by them, a serious intellectual effort has been made in the IDF to understand the nature of asymmetrical conflicts and attrition situations. As early as the late

---

60 E.g., Pedatzur, *The Arrow System*; The State's Comptroller's Annual Report No. 37 for 1986; Isaac Ben Israel, "Philosophy and Methodology of Intelligence: The Logic of Estimate Process," *Intelligence and National Security*, Vol. 4, No. 4 (October 1989), pp. 660–718; Isaac Ben Israel, "Theory of Relativity as Applied to Military Buildup," *Maarachot* 352–353 (August 1997), pp. 33–42; Dan Schueftan, "Beyond the Relative Advantage," *Maarachot* 356–357 (March 1998), pp. 70–79; Yagil Levy, "From the Citizen Army to the Market Army," in Stuart A. Cohen (ed.), *The New Citizen Armies* (New York: Routledge, 2010), pp. 196–214; Stuart A. Cohen, "The Israel Defense Forces: From a People's Army to a Professional Military: Causes and Implications," *Armed Forces & Society*, Vol. 21, No. 2 (Winter 1995), pp. 237–54.

61 Avi Kober, "From Heroic to Post-Heroic Warfare: Israel's Way of War in Asymmetrical Conflicts," *Armed Forces & Society*, Vol. 41, No. 1 (January 2015), pp. 96–122. First published online on 1 August 2013.

1990s-early 2000s, the General Staff's Training and Doctrine Department issued dozens of publications on the so-called Limited Confrontation – the IDF's new term for LIC – after having used the term "current security" for many years.

In 2001 the department issued a publication named *The Limited Confrontation*. Although it did offer some insights that might have enriched IDF commanders' minds, it was of an historical and conceptual nature, and was not formulated as a doctrine. The author, the late Colonel Shmuel (Semo) Nir, presented the limited confrontation as a non-war situation, in which the central concept was "wearing the enemy out." It was a mixture of old guerrilla warfare principles and post-modern concepts, such as the American Operations Other than War (OOTW). The *Limited Confrontation* distinguished between what was required of the military in LIC as compared to HIC situations, stressing the attritional nature of the limited confrontation, in contrast to the decisive nature expected of the military in HICs.[62] In 2004, after having studied the challenge of limited confrontation for two years, the Training and Doctrine Department started issuing "doctrinal" papers. The Infantry's "current security" doctrine department was also active in this field, but its products focused on lesson learning rather than doctrine.[63]

As part of the intellectual effort to understand the phenomenon of asymmetrical wars and the central role played by attrition in such wars, in January 2002 the National Defense College and Haifa University's National Security Studies Center dedicated a symposium to the theoretical and practical aspects of attrition in LICs.[64] In October 2004, *Maarachot* publishing house issued an edited volume titled *Limited Confrontation*, to which key Israeli military practitioners and researchers contributed articles. In many of the articles, Limited Confrontation was presented as an asymmetrical conflict, in which the vitality of interests and consequently the societal perseverance are of the highest importance, to the point of balancing military-technological capability. Some contributors characterized asymmetrical conflicts as highly affected by moral factors and pointed to the potential of manipulating the opponent's mind.[65]

62 *The Limited Confrontation* (Tel Aviv: IDF Training and Doctrine Department, 2001) [Hebrew], p. 44.

63 Erez Wiener, "From Embarrassment to Awakening," *Maarachot*, 409–410 (December 2006), p. 5.

64 Symposium on *Attrition Strategy in a Limited Confrontation*, The National Defense College, 10 January 2002.

65 Shmuel Nir, "The Nature of the Limited Confrontation," in Haggai Golan and Shaul Shai (eds), *Limited Confrontation* (Tel Aviv: Maarachot, 2004) [Hebrew], pp. 19–44; Eado Hecht, "Limited Confrontation: A Few General Features of a Unique Form of Warfare," Ibid., pp. 45–68.

TABLE 2.7 *Types of conflict per period (percentages except for N)*

	1948–1956	1957–1967	1968–1973	1974–1977	1978–1982	1983–1987	1988–1994	1995–2000	2001–2004	2005–2008
N	265	563	544	241	389	393	372	290	247	281
Unconventional	5	7.5	3	1.5	1.5	1.5	1	0.5	5	3
Conventional	84	89	93.5	97.5	97	97.5	98	95.5	69	68
Sub-conventional	11	3.5	3.5	1	1.5	1	1	4	26	29
Total	100	100	100	100	100	100	100	100	100	100

The time dimension was presented in the volume as one that requires flexible adaptation to changes in the adversaries' war objectives. Limited Confrontation was treated as a multi-dimensional effort, which has psychological, economic, physical and other aspects and is waged across all levels-of-war.[66] The volume also reflects awareness of the difficulty to achieve a decisive victory in Limited Confronations.

One of the articles, authored by Colonel (retired) Yehuda Vagman, challenged this new thinking, criticizing Israeli tendency to portray attrition as a non-military, other-than-war confrontation, and arguing that learning from the experience of other countries engaged in LICs (e.g., Indochina, Malaya, Algeria, Cuba, Zimbabwe, North Ireland) was a mistake, as their armies operated under different conditions. The author argued that the response to terror-based attrition has proven to be woefully inadequate. He called for readopting the traditional commitment to military confrontation, short war, and quick and decisive victory in order to thwart the enemy's advantages entailed in the strategy of attrition, particularly its negative psychological and economic effect.[67]

Attrition has also been referred to in *Maarachot* articles of recent years (from 2000 to 2008), many of which – between 26 and 29 percent in that period – were dedicated to LICs (see Table 2.7). Yet new approaches to LICs, which refer to attrition, such as Fourth Generation Warfare, Complex Irregular Warfare or Hybrid War,[68] have hardly been mentioned.

A More Critical Attitude toward American Concepts after 2006

In the wake of the IDF's poor performance during the 2006 Second Lebanon War, which was attributed, at least partially, to its American inspired operational conception, the IDF became more critical of American doctrine. First, it started stressing the differences between the American and Israeli challenges. Works by Dado Center researchers have argued that not only do US forces train

66 Nir, "The Nature of the Limited Confrontation."

67 Yehuda Vagman, "The Failure of 'Limited Confrontation,'" in Golan and Shai (eds), *Limited Confronation*, pp. 251–98. See also Yehuda Vagman, "Israel's Security Doctrine and the Trap of 'Limited Conflict,'" *Jerusalem Viewpoints* 514 (2004) <http://www.jcpa.org/jl/vp514.htm>.

68 Thomas X. Hammes, "War Evolves into the Fourth Generation," *Contemporary Security Policy*, Vol. 26, No. 2 (August 2005), pp. 254–63; Frank G. Hoffman, "Complex Irregular Warfare: The Next Revolution in Military Affairs," *Orbis*, Vol. 50, No. 3 (Summer 2006), pp. 395–411; James N. Mattis and Frank G. Hoffman, "Future Warfare: The Rise of Hybrid Wars," *Proceedings* 132 (2005), pp. 18–19; Frank Hoffman, "How Marines are Preparing for Hybrid Wars," *Armed Forces Journal* <http://www.armedforcesjournal.com/2006/03/1813952/ >.

for joint operations overseas, as was mentioned above, they also prepare for diverse confrontations, often fighting as part of a military coalition, whereas most of the IDF's confrontations are more similar to one another, and are usually waged on the country's borders. Moreover, the IDF has to defend the Israeli civilian rear by using multi-layer active defense systems, alongside deterrence by punishment, a challenge that the US does not face.

What the two armies still have in common is that in recent decades (the US since 2003, the IDF in the wake of the 1973 October War) their main challenge has been of LIC nature, and this shared experience has been a basis upon which the two militaries have exchanged lessons, particularly tactical ones.[69] Beyond that the particular LIC challenges of the two militaries have been different. For the US, the main LIC challenge has been COIN. According to the US Government Counterinsurgency Guide of January 2009, "counterinsurgency is the blend of comprehensive civilian and military efforts designed to simultaneously contain insurgency and address its root causes. Unlike conventional warfare, non-military means are often the most effective elements, with military forces playing an enabling role. COIN is an extremely complex undertaking, which demands of policy makers a detailed understanding of their own specialist field, but also a broad knowledge of a wide variety of related disciplines. COIN approaches must be adaptable and agile. Strategies will usually be focused primarily on the population rather than the enemy and will seek to reinforce the legitimacy of the affected government while reducing insurgent influence. This can often only be achieved in concert with political reform to improve the quality of governance and address underlying grievances, many of which may be legitimate. Since US COIN campaigns will normally involve engagement in support of a foreign government (either independently or as part of a coalition), success will often depend on the willingness of that government to undertake the necessary political changes. However great its know-how and enthusiasm, an outside actor can never fully compensate for lack of will, incapacity or counter-productive behavior on the part of the supported government."[70]

Israel's LIC thinking, on the other hand, has shifted its focus to inflicting heavy destruction on its nonstate enemies' infrastructure in order to deter

69 See, for example, Meir Finkel and Eitan Shamir. "From Whom Does the IDF Need to Learn?" *Maarachot* 433 (October 2010), pp. 28–35; Shamir, "When did a Big Mac Become Better than a Falafel?"

70 *The US Government Counterinsurgency Guide January 2009* <http://www.state.gov/documents/organization/119629.pdf>.

them from initiating new rounds of violence in the near future. In late 2008 Northern Command Chief General Gadi Eisenkot declared that in the next war the IDF would reapply the so-called "Dahiyya doctrine," named after the heavy destruction inflicted by the IAF on the Shiite quarter in Beirut during the Second Lebanon War. Eisenkot said that in the next war it will be applied with even greater destructive force and without hesitation to use disproportional firepower against civilian targets in Lebanon.[71] A few years later the IDF adopted a similar policy that has gained the name of "deterrent operations,"[72] alongside complementary operations called "mowing the grass operations,"[73] and "campaigns between the wars," in the framework of the latter clandestine activities are carried out, such as intercepting or destroying arms deliveries to hostile players, targeted killings, disrupting nuclearization processes, etc.[74]

These differences, however, have not changed the fact that both American and Israeli LIC thinking has stuck to heavy reliance on firepower in general and airpower in particular; has abandoned the traditional commitment to achieving battlefield success; and has been deeply affected by post-heroic considerations.

Climbing the Promotion Ladder from Junior to Lower-Senior Ranks Has Only Strengthened Officers' Interest in Writing

Contrary to Norman Dixon's claim that many years spent gaining promotion accentuate characteristics necessary for bureaucratic harmony but alien to intellectual activities, the *Maarachot* survey shows that as IDF officers climbed the promotion ladder from junior to lower-senior ranks their inclination to write rather increased (see Table 2.8). This finding also refutes complaints by senior Israeli officers, such as Generals Israel Tal, Amiram Levin or Uri Sagi, about suppressed criticism and too much conformity on the part of the Israeli officer corps so as to ensure promotion.[75] And indeed, lower-senior rank offi-

71 *Haarerz*, 5 October 2008.

72 Dan Har'el, "The Change in Israel's Security," Lecture at the Interdisciplinary College, 17 March 2003 <http://idclawreview.org/2013/03/17/militraylimitations2012-pt1>.

73 Efraim Inbar and Eitan Shamir, "Mowing the Grass," *Journal of Strategic Studies*," Vol. 37, No. 1 (February 2014), pp. 65–90.

74 Shay Shabtay, "The Campaign between the Wars," Maarachot 445 (October 2011), pp. 24–7.

75 Israel Tal, cited in Emanuel Wald, The *Curse of the Broken Vessels* (Tel Aviv: Shocken, 1987) [Hebrew], pp. 138–9; General Amiram Levin, describing what he witnessed as a member of the GHQ, in Amira Lam, "The General Who Knew When to Retire," *Yediot Aharonot – 7 Days Supplement*, 17 July 1998; An interview with General (retired) Uri Sagi, *Yediot Aharonot – 7 Days Supplement*, 15 May 1998.

TABLE 2.8 *Ranks in 1948-2008 in numbers and percentages*

	N	%
Lieutenant-General	32	1
Major-General	117	3
Brigadier-General	128	4
Colonel	360	10
Lieutenant-Colonel	339	9
Major	208	6
Captain	77	2
Lieutenant	36	1
NCO	8	0
Non-regulars/Anonymous	2280	64
Total	3585	100

cers are usually more experienced and educated than junior officers, and seem to feel fairly confident and authoritative, personally as well as professionally.

The Detrimental Effect of the IDF's Lack of Intellectualism

The Cost of Poor Intellectualism

Although it is difficult to draw a clear line between little appreciation for history-based military theory and the IDF's quality and performance on the battlefield, such correlation does seem to exist. Four examples can illustrate the extent to which the neglect of the intellectual aspects of the military profession affected the IDF's operational performance. The first example is the failure to implement the doctrinal and organizational requirements of mountainous terrain warfare's principles against Syrian troops on the central front of the 1982 First Lebanon War, which constituted the main cause for the IDF's difficulties on that front. The IDF's poor performance was apparently unexpected, as it enjoyed quantitative superiority and its operational planners were forewarned months in advance that they might be fighting in that particular terrain. In May 1981, a year before the war broke out, a special issue of *Tziklon* was dedicated to mountain warfare.[76] The anthology covered theoretical

76 *Tziklon*, No. 8 (May 1981).

aspects of such warfare, and included valuable data on Soviet and Arab doctrines, Soviet experience in Afghanistan, and British experience in fighting in Syria and Lebanon during World War II. Had Israeli military planners read *Tziklon*, the difficulties they faced might have been avoided. From the experience of others, particularly from the British experience in the 1941 Operation Exporter against Vichy troops, they would have learned that mountain warfare ought to be an infantry-based, wide front, and indirect operation, principles that they have violated.

It should be noted, though, that the Israeli failure was not total. A paratroopers brigade headed by a talented commander, Colonel Yoram Yair, conducted classical mountain warfare east of Beirut. Unlike many commanders in the IDF, Yair *was* familiar with the principles of mountain warfare, and prior to the war had trained his brigade for fighting in such terrain.[77] The paratroopers under his command first outflanked the enemy by landing on the seashore near the outlet of the Awali River, and then by advancing from southwest to northeast through the mountains encircling Beirut, bypassing the seashore axis from the Israeli border north of Beirut. Despite strong Syrian and Palestinian resistance – both in the mountains or in the streets of Lebanese towns and villages along the critical passes – the paratroopers handled the challenges efficiently, thanks to their flexibility as infantry. They seized necessary passes, then opened lines of operation and logistics for the Israeli armor, and successfully secured the passes. They eventually managed to take control of the Syrian outer defense belt around Beirut and to join up with the Christian-Lebanese allies north of the city.[78] The excellent performance and efficiency of the paratroopers in this case only underlined the general military misfortune against the Syrians on the central front.

The second example pertains to popular uprising. Had the IDF studied the events of the 1936–39 Arab Revolt (which is sometimes referred to as the First Intifada) and the Gandhi-led civil disobedience during India's struggle for independence between the 1920s and the 1940s, it might have been better prepared for the 1987 Intifada, particularly during its early stages. The IDF would have learned from the revolt first, that popular uprisings can erupt spontaneously as a result of a nationalistic impulse and a growing sense of frustration, only later being exploited by the leadership for a more organized struggle. Second, that the economic dependence of a populace on the ruler can make civil uprising very destructive for the people involved in it, in the long run. The

77 Yoram Yair, *With Me From Lebanon* (Tel-Aviv: Maarachot, 1990) [Hebrew], p. 14.

78 Benny Mem, "The Peace for Galilee War: Main Operations," *Maarachot* 284 (September 1982), pp. 24–48.

six-month period of civil disobedience during the revolt came to an end as a result of merchants' pressures, once Palestinian economic life verged on collapse. Had the Israelis learned this lesson they would have refrained from making ongoing attempts to open Palestinian shops during the 1987 Intifada. Too much time was wasted before they came to the conclusion that closures could deprive many Palestinian families of their income and break their will to continue the struggle. Third, the Israelis could have learned from the Arab Revolt that quashing an uprising cannot be easily translated into political achievements, if at all. Despite British achievements in dealing with the violence produced by Arab guerrillas and terrorists and in wearing down Arab civil disobedience, the Arab revolt brought about political achievement for the Palestinians, particularly MacDonald's White Paper, which imposed restrictions on Jewish immigration to Palestine, land purchase by Jews in Palestine, and Jewish settlement there. The Israelis should not have been surprised, then, that despite the fact that the Intifada failed in the narrow sense, the Palestinians reaped political fruits in terms of the opening of a dialogue with the US and later on with Israel. In light of the similarities between the two cases and the fact that both Intifadas occurred on the very same scene and in the framework of the same conflict, the Israeli failure to learn from the revolt can certainly be considered a major flaw.

The third example refers to one of the peaks of Israeli military incompetence – the Second Lebanon War. A close look into the IDF's failures reveals that they were highly shaped by a lack of theoretical and empirical knowledge. A few examples will demonstrate how knowledge could have prevented IDF incompetence during that war. Had the IDF been acquainted with the relatively short history of airpower, they would have known that no victory at the strategic level has ever been achieved from the air. Kosovo, which was so frequently referred to as a model of victory from the air, was a grand-strategic victory, achieved by denying Serbian *society* the ability to carry on the war – not the Serbian army, which remained almost unharmed. But IDF planners, particularly Chief-of Staff General Dan Halutz (formerly IAF Chief), were so confident that airpower alone – or almost alone – could do the job,[79] that they did not provide the government with any real alternative plan until the latest stage of the war. It is ironic that two Israeli airpower experts – Generals Amos Yadlin and Ido Nehushtan – eventually understood this during the

79 Halutz's testimony before the Winograd Commission <http://www.vaadatwino.org.il/pdf/תמליל%20 דן%20 חלוץ.pdf>, p. 16.

Second Lebanon War, whereas the chief-of-staff – an airman himself – failed to acknowledge this.[80]

The fourth example relates to logistics. Had IDF officials in charge of logistics read Martin Van Creveld's writings on technology and logistics, they might have internalized the fact that unlike a nonmilitary context, in which the preferred logistical system is a centralized one, as it allows better control over logistical resources and saves manpower and stocks, in an operational context a centralized system may cripple the combat units' logistical autonomy. Understanding the difference would have precluded the IDF from operating without an effective logistical tail, as happened during the First and the Second Lebanon Wars.[81]

It is reasonable to believe that proficiency, not only in the practical aspects of the military profession but also in its abstract aspects, helps create professional pride among commanders and improves their image in the eyes of the society, and can be translated into strengthened self-confidence of commanders in their profession, and vice versa. A small-scale research conducted by Colonel Amir Abulafia, during his studies at the IDF's National Defense College, investigated the courage of IDF commanders to express independent opinions. For that purpose, Abulafia interviewed eight senior commanders holding the ranks of Major-General and Brigadier-General, both in active service and reserves, asking for their opinion, based on their long experience. Most of them (62 percent) believed that lack of acquaintance with basic professional knowledge had often been translated into lack of courage to differ with superiors, and the adoption of mainstream conceptions.[82]

The Danger Entailed in Emulated Doctrines

Learning from foreign armies' doctrines is conducive as long as it is made in a selective manner, and doctrines are not adopted irrespective of the difference in the strategic conditions. Emulating doctrines would be justified, if at all,

80 In a publication of the Institute for Air and Space Strategic Studies, former IAF chief and former Deputy Chief-of-Staff David Ivry wrote that air power could not be victorious by itself in the war against terrorism. Zeev Schiff, "The Foresight Saga," <www.haaretz.com/hasen/spages/749268.html>. And IAF Chief Eliezer Shkedy warned that nobody should expect the IAF to stop the Katyusha fire. "Expect a success of no more than one to three percent in hitting the Katyushas," he said. Shelah and Limor, *Captives in Lebanon*, p. 138.

81 Alon Ben-David, "Israel Introspective after the Lebanon Offensive," *Jane's Defense Weekly*, 22 (August 2006), p. 19.

82 Amir Abulafia, "The Courage to Express Independent Opinions," *Maarachot* 433 (October 2010), p. 23.

only if the strategic conditions and the challenges of the armies concerned were similar.

Throughout the years, up until the 1990s, the IDF's doctrinal thinking was influenced by the doctrines of foreign armies only to a limited degree. A change occurred in the 1990s, when the IDF became deeply fascinated with American RMA ideas. What caught its imagination in particular was the combination of intelligence, surveillance, and reconnaissance; advanced command, control, communications, computers, and intelligence, which – alongside precision strike capability –promised rapid, decisive victory, fewer casualties, and little collateral damage.[83]

The IDF launched its most RMA-oriented operational doctrine in April 2006, three months before the outbreak of the Second Lebanon War. An operational order issued by the General Staff on 13 July 2006 described the upcoming operation as a protracted offensive, based on the combination of accurate, standoff fire and limited operations on the ground.[84] The new, emulated doctrine, however, soon proved its inadequacy.[85] This should not have come as a surprise, given the differences in characteristics and strategic conditions between the Israeli and the American armies. The latter is much bigger, has huge resources at its disposal, and enjoys a much greater international freedom of action. It often operates in alliance with its NATO allies, it is a volunteer-based military, it is structured differently than the IDF, and its society is hardly under direct threat. There were areas, though, such as LICs, where the IDF was more experienced than the American army, at least up to the end of the Cold War. At the same time, it could have benefitted from the American commanders' education and training programs (see discussion of this in Chapter 5).

Damage Inflicted by Phony Intellectualism

OTRI – the Operational Theory Research Institute – which was established by the IDF's Doctrine Branch in 1995, was supposed to develop operational art knowledge that would be useful to commanders at the operational level. It is hard to evaluate the contribution of OTRI in light of the fact that even after

83 Elizabeth A. Stanley-Mitchell, "The Digital Battlefield: What Army Transformation Efforts Say about its Professional Jurisdiction," in Don M. Snider and Gayle Watkins (eds), *The Future of the Army Profession* (New York: McGraw Hill, 2003), pp. 155–78; Benjamin S. Lambeth, *The Transformation of American Air Power* (Ithaca: Cornell University Press, 2000), p. 303.

84 Alex Fishman, "The Five-Day War," *Yediot Aharonot Weekend Supplement*, 17 April 2007.

85 Avi Kober, "The IDF in the Second Lebanon War: Why the Poor Performance?" *Journal of Strategic Studies*, Vol. 31, No. 1 (February 2008), pp. 3–40; Isaac Ben-Israel, *The First Missile War* (Tel Aviv: The Security Studies Program, May 2007) [Hebrew].

years of activity it failed to produce written materials. But is seems quite certain that it inflicted damage at least in one important area, namely, its use of flowery "neologisms" by the institute's instructors, which the Winograd Commission's final report on the Second Lebanon War described as potentially creating a "tower of Babel" that impeded the creation of common language and mutual understanding among commanders, particularly between commanders at the operational and tactical levels. Indeed, according to the Commission's report, this language was considered by many IDF commanders to be obscure, unclear, confusing, and empty, and the combination of a problematic operational conception and unclear orders was among the factors that accounted for the IDF's poor performance on the battlefield.[86] This especially characterized the 91st Division, whose commander was one of the staunchest believers in the new language.[87]

Rand expert Russell W. Glenn called the flaws in the IDF's operational conception an "intellectual virus."

> The intellectual virus that many in the IDF fear has infiltrated their military's thinking has both domestic and international roots. Israel's own theorists seem to have overlooked the need to ensure that these ideas were accessible to those whom the armed forces must train. Imported concepts, such as effect-based operations, came under attack as having failed to meet the test of combat conditions. There is a need to recognize the inherent value of simplicity and clear prose when writing doctrine and developing ideas that ultimately will influence the men and women actually confronting real-world challenges.[88]

Even after the Winograd Commission's report was completed and made public, one of OTRI's greatest proponents, the Chief of the Military Colleges, was still in office, preaching ideas that were rejected by the Commission and by the IDF itself, inspired by post-modern philosophers such as Gilles Deleuze, Félix Guattari, and architects, rather than military thinkers.[89]

86 Ibid., pp. 274–5, 321.

87 The Winograd Commission's Report <http://www.vaadatwino.org.il/pdf/דוח%20סופי.pdf>, pp. 318, 322.

88 Russell W. Glenn, *All Glory Is Fleeting Insights from the Second Lebanon War* (Santa Monica: Rand, 2012), p, XIV.

89 Gershon HaCohen, "Designing the Space and a Military Campaign during the Disengagement [from Gaza]," *Maarachot* 432 (August 2010), pp. 24–34.

Modern Focus

Israeli military thinking has been slow in adapting or reacting to changes in the nature of war, and its sluggishness at times enabled the development of conceptions that were later on contested by reality.

Types of Conflict

In light of the dominance and pervasiveness of sub-conventional conflicts, one would expect the IDF to be more committed to the study and analysis of the nature of LICs and their implications. And indeed, during Israel's early years, "current security" challenges along the borders, particularly *fedayeen* raids, triggered interest in LICs, which also found expression in a greater number of *Maarachot* articles dedicated to such types of conflicts. Between the mid-1950s and the late 1980s, however, the IDF focused on the challenge of conventional wars with Arab states (see Table 2.7). Explanations for this should be sought not only at the systemic level but also at the unit (state) level, and are referred to in Chapters 3 and 4. In later periods, particularly during the years 2000–2008, authors of *Maarachot* "rediscovered" LIC challenges, and this is reflected in the growing number of articles dedicated to LIC-oriented issues – between 26 and 29 percent of the articles during that period (see Table 2.7).

Only 3 percent of the articles published in *Maarachot* in the period 1948–2008 dealt with the challenge of unconventional war (see Table 2.9).

This does not necessarily reflect lack of interest in the topic but should rather be attributed to the Israeli government's censorship on any discussion of Israeli nuclear capability on formal outlets such as *Maarachot*. An exception in this respect could be considered the fact that between 1948 and 1967, articles on nuclear weapons were published, although not with reference to Israel but rather to the general impact of nuclear weapons on global strategic stability. During these years, the percentages were relatively high – 5 percent in 1948–1956, and 7.5 percent in 1957–1967. Heightened interest can be traced also after the year 2000 – 5 percent in the period 2000–2004, and 3 percent in the period 2005–2008 (see Table 2.7) – which most probably reflected the growing anxiety about Iran's nuclear project.

It is worth noting that most of the articles on biological and chemical weapons were published between the years 1968–1973, probably reflecting, at least indirectly, an interest in the Arab tendency to compensate for Israel's perceived conventional ascendancy and nuclear monopoly by acquiring such "equalizers."

TABLE 2.9 *Types of conflicts 1948-2008 in numbers and percentages*

	N	%
Unconventional	112	3
Conventional	3232	90
Sub-conventional	241	7
Total	3585	100

Levels-of-War

As already argued in Chapter 1, whereas "regular" war is usually waged across the entire range of the levels-of-war pyramid, in LICs in particular military encounters often take place at the tactical level, and are usually limited in terms of forces, time, or place. On the other hand, the objectives of those engaged in such conflicts and sometimes also their targets are beyond the direct battlefield, at the grand-strategic level, namely, the enemy's society and economy.

Israel has experienced such a reality, particularly since the late 1970s-early 1980s, both in the territories and on its borders with Lebanon and Gaza. This, however, is not reflected in the *Maarachot* survey, which portrays a relatively balanced picture, with 8 percent of the total number of *Maarachot* articles in the period 1948–2008 dealing with tactical matters, 3 percent with the operational level, and strategy and grand-strategy with 4 percent each (Table 2.10).

Interesting in particular is the fact that in the period 1948–1967, a significant number of levels-of-war related articles – 40 percent in the period 1948–1956, and 13 percent between 1957 and 1967 – focused on tactical issue. Only in the late 1980s/early 1990s did the importance of the operational level start receiving greater recognition, peaking at 7 percent in the 2000s, and jumping into the first place. Finally, since the late 1980s, greater attention has been paid to the grand-strategic level (see Table 2.11).

This distribution indicates, first, that the IDF kept preparing itself for confrontations of a HIC nature. Second, that the IDF has had a strong tactical orientation, the reasons for which seem to be twofold: it is easier to think narrowly; and commanders are mostly affected by their tactical experience and training. Third, that the surge in the interest in the operational level seems to have occurred as a result of the impact of OTRI. Fourth, that the rising interest in the grand-strategic level must have been affected by the outbreak of the Intifadas. Commanders at the tactical level have become more sensitive to the non-military aspects and repercussions of their military activity, incorporating

TABLE 2.10 *Levels of war 1948-2008 in numbers and percentages*

	N	%
Grand Strategy	130	4
Strategy	161	4
Operational level	95	3
Tactics	299	8
No reference to levels of war	2900	81
Total	3585	100

such considerations, which are often of grand-strategic nature, in their tactic-related decisions. Commanders engaged in LICs have become soldier-statesmen rather than combat leaders. This has accounted for the "grand-strategization of tactics" phenomenon.[90] Illustration of such command-and-control reality is the summation by Prime Minister Ariel Sharon of his meeting with a group of IDF colonels engaged in LIC in the West Bank and the Gaza Strip, in early 2002: "As a young officer, whenever I met with politicians, I spoke tactics, and they spoke strategy. With you, I speak tactics, while you speak strategy," he complained.[91]

In recent decades critics of the IDF have claimed that its education and training system has limited itself to the tactical level.[92] In 2007, the Knesset Foreign Affairs and Defense Committee released a report that pointed to the lack of an educational track bridging between the tactical level (i.e., the Tactical Command College and the Command and Staff College), on the one hand, and what the Committee referred to as the "strategic level," i.e., the National Defense College, on the other.[93] A sub-committee that explored the matter in greater depth warned of the lacunae in senior commanders' education above

90 Bernard Boëne, "Trends in the Political Control of Post-Cold War Armed Forces," in Stuart Cohen (ed.), *Democratic Societies and Their Armed Forces: Israel in Comparative Context* (London: Frank Cass, 2000), pp. 73–88; Eliot A. Cohen, "Technology and Supreme Command," Ibid., pp. 89–103.

91 *Yediot Aharonot*, 1 February 2002.

92 Eytan Gilboa, "Educating Israeli officers in the Process of Peacemaking in the Middle East Conflict," *Journal of Peace Research*, Vol. 16, No. 2 (1979), pp. 155–62; Harkabi, *War and Strategy*, pp. 587–8; Handel, *Masters of War*, pp. 353–60; Tamari, "Is the IDF Capable of Changing in the Wake of the Second Lebanon War?", pp. 26–41.

93 The Committee's report is available at <http://my.ynet.co.il/pic/news/3.7.2007/hub.doc>.

TABLE 2.11 *Levels-of-war per period (percentages except for N)*

	1948–1956	1957–1967	1968–1973	1974–1977	1978–1982	1983–1987	1988–1994	1995–2000	2001–2004	2005–2008
N	265	563	544	241	389	393	372	290	247	281
Grand Strategy	3.5	2	2	4	3	3.5	4.5	3.5	9	6
Strategy	5	5.5	3.5	4	4.5	3.5	6	3	5	5
Operational level	5	1	1.5	1	1	1	2.5	4.5	6	7
Tactics	40	13	5	2.5	3	3	5	5	6	6
No reference to levels of war	46.5	78.5	88	88.5	88.5	89	82	84	74	76
Total	100	100	100	100	100	100	100	100	100	100

the tactical level, which was exposed during the Second Lebanon War.[94] Ironically, the Knesset's sub-committee was not aware of the fact that military issues have hardly been studied at the National Defense College in recent years.[95]

Dimensions of Strategy

A Technological Bent

During Israel's early years, Israeli senior commanders understood the danger entailed in over-reliance on technology at the expense of the human factor.[96] In 1960, Lieutenant-Colonel (retired) Israel Beer – one of the outstanding Israeli military intellectuals at the time – published an article in *Maarachot* in which he warned against the tendency that had spread across the Western world to put too much faith in technology, and called for a balanced approach. "[Military] conceptions that concentrate too much on technology are no less conservative than military thought that ignores the developments on weapon systems."[97] Beer asserted.

In recent decades, the IDF has found great interest in the non-operational dimensions of strategy, with particular interest in technology. In the 1990s, Israeli senior commanders were still trying to develop or purchase the best weapon systems in order to ensure Israel's technological edge, but at the same time they kept a balanced approach towards technology. More recently, though, technology started overshadowing the non-material aspects of Israeli strategy and tactics, to the point of developing into a cult of technology. According to General (retired) Isaac Ben-Israel, Israel should pursue a technology-focused military doctrine, force design, and military buildup as military quality is now identified with high-tech capabilities.[98]

The *Maarachot* survey shows that the dimension of strategy most prevalent among *Maarachot* articles between 1948 and 2008 was the technological one, with 8 percent of the articles (see Table 2.12).[99]

---

94 <http://www.nrg.co.il:80/online/1/ART1/603/588.html>.

95 Testimony by General (retired) Amatzia Chen, a former instructor at the National Defense College <http://www.kav.org.il/100994/647>.

96 Reuven Gal, *A Portrait of the Israeli Soldier* (Westport: Greenwood, 1986), p. 175; Cohen et al., *Knives, Tanks, and Missiles*, pp. 63–5.

97 Israel Beer, "Conservatism and Flexibility in Military Thought," *Maarachot* 126 (1960), p. 26. See also pp. 26–8, 48.

98 Ben-Israel, "The Military Buildup's Theory of Relativity," p. 33; Isaac Ben-Israel, "Security, Technology, and Future Battlefield," in Golan (ed.), *Israel's Security Web*, p. 279.

99 Kober, "Israeli Military Thinking as Reflected in *Maarachot* Articles, 1948–2000," p. 156.

TABLE 2.12 *Dimensions of strategy 1948-2008 in numbers and percentages*

	N	%
Operational	144	4
Societal	82	2
Technological	309	8
Logistical	107	3
No reference to dimensions of strategy	2943	82
Total	3585	100

The Decline of the Operational Dimension: From Decisive Victory to "Victory Image"

Another phenomenon that has characterized the operational dimension in recent years has been the weakening of the IDF's commitment to clear-cut military achievements on the battlefield, both in general and in LIC contexts in particular. In 2001 Chief of the IAF Dan Halutz said that the physical aspect of battlefield decision had lost its importance.[100] In an interview with Brigadier Eyval Gil'adi from IDF's Planning Branch, held prior to his retirement from a military career, the senior officer said: "When I started my job, I found in the plans the term, 'defeating the Palestinians.' I asked myself, what is that nonsense? Whom exactly are we supposed to defeat? What does defeat mean? We tried to think of alternatives to defeating the enemy. Initially I talked about a 'victory image,' which is merely an appearance. It then became a matter of producing a victory show."[101] Referring to the Battle of Tul Karm in early 2002, Colonel Moshe (Chico) Tamir said: "We wanted pictures of armed Palestinians raising their hands in a gesture of surrender to be broadcast. We learned from Hezbollah that the consciousness battle was not less important than a real victory on the battlefield."[102] A few years later, during the Second Lebanon War, Chief-of-Staff Dan Halutz ordered the paratroopers brigade to enter the town of Bint Jbeil in southern Lebanon in order to hoist the Israeli flag over the former IDF's headquarters building in that town and take photos of it. This was supposed to create a "victory picture" that would substitute the need to

100 Dan Halutz, Lecture at the National Defense College, 28 January 2001.

101 *Yediot Aharonot Weekend Supplement*, 19 September 2003.

102 Amos Harel, "Educating Chico," *Haaretz*, 1 August 2005 <http://www.haaretz.co.il/misc/1.1558854>.

capture the town physically. Subsequent to hoisting the flag, the paratroopers' commander was supposed to deliver a victory speech there and a small military parade was supposed to follow, in order to demonstrate the Israeli "victory."[103]

General Moshe Yaalon, too, expressed skepticism about the ability to land a decisive blow on a guerrilla organization like Hezbollah.[104] Already as Commander of the Central Command he came to the conclusion that "battlefield decision" and "victory" were irrelevant in a reality of asymmetrical, low-intensity conflicts. He believed that the classical military thought had to be replaced by an operational language that would fit an era where the media and consciousness became the most substantial factors.[105] No wonder that when the Second Lebanon War broke out, Chief of Operations General Gad Eisenkot said that defeating Hezbollah was unattainable.[106] "The military did not even pretend to achieve decisive victory on the battlefield," was Foreign Minister Tzipi Livni's impression from the IDF's state of mind during a Cabinet meeting held on July 31, 2006.[107] And in September 2009, the Chief of the IDF's Planning Department, General Amir Eshel, said: "We still think in terms that are no longer relevant to the war of today, which are borrowed from conventional war. One cannot eliminate a terror organization, which is deeply rooted in the hearts of its people. One can only hit and deter it."[108]

The Erosion of Traditional Force Multipliers

As is pointed out in Chapter 4, the IDF's quantitative inferiority vis-à-vis a coalition of enemies, as well as the small size of Israel and lack of strategic depth, dictated operational art that focused on compensating for weaknesses via a series of force multipliers. Once technology in general and the ascendancy of firepower over maneuver in particular impinged upon the IDF's force buildup and military operations, Israeli maneuver-oriented force multipliers were the first to pay the price.

---

103 Ibid., p. 192.

104 Ibid., p. 129.

105 Tamari and Klifi, "The IDF's Operational Conception," pp. 33–5.

106 Ibid., p. 54.

107 Ariella Ringel-Hoffman, "This Is Not How a War Should Be Conducted," *Yediot Aharonot Weekend Supplement*, 23 March 2007.

108 *Haaretz*, 26 September 2009.

A Changing Logistical Logic

During the mid-1960s, the IDF underwent a major logistical reform. Divisions or brigades were directly "pushing" supplies to their own forces along the lines of operation.[109] The conditions on the battlefield – the ascendancy of maneuver, high motivation, relatively short lines of communication, and personal acquaintances between the providers of supplies and the fighting forces – made such a decentralized system feasible.

The ascendancy of firepower over maneuver imposed on the IDF a logistical reorganization in the direction of a more centralized system, based on modularly structured area-logistics units.[110] During the Second Lebanon War, the new conception was put to test and exposed its weaknesses. It may have provided a better control over logistical resources and may have saved manpower and stocks,[111] but, at the same time, it crippled the combat units' logistical autonomy and countered operational art's logic and spirit. Had the war involved large-scale ground maneuvers, it is doubtful that it would have met operational requirements. It is no wonder, therefore, that the IDF rethought this and decided to strengthen the logistical autonomy of the field units at the division level.[112]

Post-Heroic Mindset and Moral and Legal Considerations Becoming an Integral Part of Israeli Military Thinking

Some of the previous sections of this chapter have referred to post-modern inputs that permeated Israeli military thinking during the 1990s and the early 2000s, describing them as "phony intellectualism." But there has been at least one major significant aspect of post-modernism that does not deserve to be treated as a phony intellectualism, which is post-heroic warfare. Post-heroic warfare has two major rules that constitute a basic, substantial deviation from previous centuries, during which war had been a unique social phenomenon involving killing and getting killed. The first rule dictates that one is not "allowed" to get killed, whereas the second rule is that one is also not allowed to kill innocent enemy civilians. Post-heroic warfare is compatible with America's relatively new way of war, which is characterized by the belief in technological superiority, air dominance, excellent intelligence, and accurate

---

109 Luttwak and Horowitz, *The Israeli Army*, p. 175.

110 Kober, "The IDF in the Second Lebanon War," p. 29.

111 Amnon Barzilai, "[Chief of the IDF's Technology and Logistics Branch General Udi] Adam's Technological Revolution," *Haaretz*, 2 April 2004.

112 <http://dover.idf.il/IDF/News_Channels/bamahana/07/43/02.htm>.

firepower, as means of minimizing both America's own casualties and "collateral damage." [113]

The concept of post-heroic warfare was first coined by Edward Luttwak in the mid-1990s.[114] Luttwak formulated it after having noticed that in their interventions in the post-Cold War conflicts in the Gulf, the Balkans and elsewhere, the commitment of the US and its allies to avoid casualties among their own troops had sometimes come at the expense of operational effectiveness. Explanations for post-heroic warfare are offered in Chapter 4.

It should be noted, though, that a post-heroic mindset had existed in the IDF long before the concept of post-heroic warfare was explicitly defined and formulated. But like Monsieur Jourdain in Moliere's play, *The Bourgeois Gentleman*, who discovered that "for more than forty years I have been speaking prose without knowing anything about it,"[115] most IDF commanders had neither heard about it nor written about it in *Maarachot*. In the Israeli case, the strongest explanation for such behavior has been the marriage of two factors: Israel's engagement in LICs, which have not threatened its basic security, let alone its existence; and its sophisticated technology. The explanation for the Israeli post-heroic turn could not be demographic, as birth rates in Israel have been much higher than in the UK and the US: 22 births per 1,000 inhabitants, comparative to 13, and 14 for 1,000 inhabitants, respectively.[116]

Israel has always been casualty averse. As Prime Minister Ben-Gurion put it, "[We are committed to] achieving victory with minimal losses."[117] But since the late 1970s post-heroic behavior has gradually become an integral part of its military thinking, strategic culture and way of war, often coming at the expense

113 H.H. Gaffney, *The American Way of War through 2020*, The CNA Corporation <http://www.au.af.mil/au/awc/awcgate/cia/nic2020/way_of_war.pdf>; Brian M. Linn, *The Echo of Battle: The Army's Way of War* (Cambridge: Harvard University Press, 2007); Brian M. Linn, "'The American Way of War' Revisited," *The Journal of Military History*, Vol. 66, No. 2 (April 2002), pp. 501–33; John A. Lynn, *Battle: A History of Combat and Culture* (Boulder: Westview Press, 2003); Colin S. Gray, *Irregular Enemies and the Defense of Strategy: Can the American Way of War Adapt?* (Carlisle: US Army College, March 2006).

114 Edward N. Luttwak, "Where Are the Great Powers?" *Foreign Affairs*, Vol. 73, No. 4 (July/Aug. 1994), pp. 23–8; "Toward Post-Heroic Warfare," *Foreign Affairs*, Vol. 74, No. 3 (May/June 1995), pp. 109–22; "A Post-Heroic Military Policy," *Foreign Affairs*, Vol. 75, No. 4 (July/Aug. 1996), pp. 33–44; Edward Luttwak, "Post-Heroic War," *Maarachot* 374–375 (February 2001), pp. 4–9.

115 Moliere, *The Middle Class Gentleman* (*Le Bourgeois Gentilhomme*), Part 1 <http://www.fullbooks.com/The-Middle-Class-Gentleman-Le-Bourgeois1.html>.

116 <http://data.worldbank.org/indicator/SP.DYN.CBRT.IN>.

117 Ben-Gurion, *Yichud Ve-Ye'ud*, p. 13.

of mission fulfillment.[118] Reliance on firepower instead of maneuver and on air power instead of ground forces; the use of stand-off precision weapons, and drones; and the development and deployment of active defense systems – have all become means of ensuring low casualties among IDF troops. At the same time, issues of just war, discriminate use of force, proportionality, and civil liberties have become an integral part of Israeli military thought and particularly counterinsurgency policy. Israel's strong commitment to fight morally has been expressed *inter alia* by the development of doctrinal and technological means and information-gathering methods that could reduce collateral damage considerably; the existence of a code of ethics, which was formulated by the IDF as a result of the ethical dilemmas Israeli troops faced during the Intifadas; close control by the IDF's judicial authorities on targeted killing of terrorists and other operations in the territories; rules of engagement and methods of dispersing demonstrations that tried to ensure the minimization of loss of life or serious bodily injury; and occasional rules by the Israeli Supreme Court on matters such as discriminate use of force, torture, and human shields.

Since the late 1990s, military lawyers have become involved in operational aspects, something that might have subordinated operational considerations to legal ones. Although in reality the military legal system has often authorized operations and adopted relatively flexible interpretations of the law in order to justify IDF operations, the more legal advisers were involved in operational matters the greater the chances that operational considerations would be subordinated to legal ones, to the point of degenerating operational sophistication, freedom of action, and operational effectiveness. It is no wonder, therefore, that in its report on the management of the Second Lebanon War, the Winograd Commission expressed concern of the growing reliance on legal advice in the course of military operations, which it considered liable to shift the responsibility from commanding officers to advisers and to divert commanders' attention from their operational challenges[119] – a phenomenon that gained the name Judicialization.

Conclusion

Throughout the years the IDF has shown symptoms of poor intellectualism, which have been reflected, among other expressions, in a lack of interest in the

118 Kober, "From Heroic to Post-Heroic Warfare."

119 Chapter 14, paragraphs 29 and 31 of the report.

theoretical aspects of the military profession and the underestimation of theory's contribution to practice. This lack of intellectualism has had a detrimental effect on the IDF's performance, particularly during the First and Second Lebanon Wars and the First Intifada. Since the 1990s, the IDF has been emulating an RMA-inspired American doctrine, which has come at the expense of its originality and innovation. As if to add insult to injury, Israeli military thinking has been affected by false intellectualism and intellectual pretense as well.

These negative trends have been mitigated by a number of positive symptoms, such as vibrant military thinking during Israel's formative years, both before the establishment of the State of Israel and during its early years; the relative popularity of military history; the existence of great debates on operational and buildup issues; or the establishment of the Tactical Command College in the 1990s.

As far as the modern focus of Israeli military thought is concerned, this chapter has revealed four major characteristics. First, a late adaptation to LIC challenges, and a negative effect of policing missions in the territories on the IDF's military thought and performance in LIC situations. Second, a strong tactical orientation, and a late adaptation to the operational and grand-strategic levels since the 1980s-early 1990s. A two-way linkage has emerged between the tactical and the grand strategic level, which has manifested itself in the "tacticization of grand-strategy" and the "grand-strategization of tactics." Third, a too-strong technological orientation of the IDF in recent decades, which has gradually become a second nature, and has eroded the IDF's operational art. Other aspects illuminated in the chapter are the erosion of Israel's traditional force multipliers, to a great extent due to the ascendancy of firepower over maneuver; and a changing logistical logic as a result of the declining role played by maneuver. Finally, post-heroic mindset has become an integral part of Israeli military thinking.

CHAPTER 3

Systemic Formative Factors

This chapter focuses on three systemic formative factors in Israeli military thought – the pervasiveness of LICs, the technological dimension, and legal and moral constraints created by international law and norms.

The Pervasiveness of LICs

Although LIC reality has been an important part of Israeli security environment since the 1950s, for many years Israel treated LICs as a relatively minor challenge, which did not merit too much intellectual investment. LICs were referred to as a "current security" challenge, as compared to the "basic security" challenge posed by the regular armies of Arab states.[1] As Shimon Peres once put it, "referring to current security as the major security challenge [...] would be like fixing an unstitched dress when the entire body is in danger."[2] The IDF's military thinking was geared toward HICs,[3] which is reflected in officers' professional articles published in *Maarachot* between the mid-1950s and the late 1980s. Only in later periods, particularly since the mid-1990s, did *Maarachot* contributors start relating more seriously to LIC challenges (see Table 3.1).

In the 1980s, for the first time, LICs started being treated as challenging Israel's "basic security," either actually or potentially. The main explanation for this "upgrading" is the aggravated threat posed by Palestinian terrorism. Although if measured objectively, Israel's LICs have never constituted an existential threat to the country, nevertheless, senior Israeli officials started referring to them as such. In the early 1980s Defense Minister Ariel Sharon introduced new Israeli *casi belli*, which included, for the first time, a tacit *casus belli* that related to insurgency from neighboring states.[4] In 1986 Foreign

1 For the distinction between "current security" and "basic security," see Shimon Peres, *The Next Phase* (Tel Aviv: Am Hasseffer, 1965) [Hebrew], pp. 9–15.

2 Ibid., p. 11.

3 Cohen et al., *Knives, Tanks, and Missiles*, p. 71.

4 Zvi Lanir (ed.), *War by Choice* (Tel Aviv: Jaffee Center for Strategic Studiese, 1985) [Hebrew], pp. 157–63.

 | DOI 10.1163/9789004306868_004

TABLE 3.1 *Types of conflict per period (percentages except for N)*

	1948–1956	1957–1967	1968–1973	1974–1977	1978–1982	1983–1987	1988–1994	1994–2000	2001–2004	2005–2008
N	265	563	544	241	389	393	372	290	247	281
Unconventional	5	7.5	3	1.5	1.5	1.5	1	0.5	5	3
Conventional	84	89	93.5	97.5	97	97.5	98	95.5	69	68
Sub-conventional	11	3.5	3.5	1	1.5	1	1	4	26	29
Total	100	100	100	100	100	100	100	100	100	100

Minister Yitzhak Shamir reacted to an attempt made by a Syrian-dispatched Abu Nidal operative to place a bomb on an El Al plane leaving London's Heathrow Airport for Tel Aviv[5] by declaring that had the aircraft exploded, Israel might have launched a war against Syria.[6]

Four months after the outbreak of the 1987 First Intifada, Defense Minister Yitzhak Rabin realized that Israel's military might was almost irrelevant to coping with extremely limited violence or civil disobedience, and that Israel's status as the source of power and authority in the territories dictated that its behavior there be restricted both legally and politically. He also became aware that the challenge could not be surmounted in one attempt but rather through a cumulative process of exhaustion,[7] and that the solution for the underlying causes of the uprising was political.[8] This constituted a watershed in Israeli security conception.

In the 1990s and 2000s, Israeli leadership continued relating to LICs in terms of a basic security threat. In 1995, Prime Minister Yitzhak Rabin declared that for Israel terrorism represented a "strategic threat."[9] And in the midst of a wave of murderous terrorist attacks by Palestinian suicide bombers against Israeli citizens during the Second Intifada, which erupted in 2000, Chief-of-Staff Shaul Mofaz said that for Israel the conflict was "an existential war,"[10] while Prime Minister Sharon declared that the 2002 Operation Defensive Shield against Palestinian terrorism was "over our home."[11]

At the same time, a belief spread among IDF senior commanders that asymmetrical wars became "wars of conviction," or "wars of consciousness," which meant that it was no longer necessary to defeat the enemy in the traditional way but rather to create a victory appearance (see discussion of this aspect in Chapter 2). Although one of the lessons learned from an exercise (Firestones-9) carried out two years before the outbreak of the Second Lebanon War was that in order to stop the launching of rockets onto Israeli territory it was necessary to affect the enemy's capabilities rather than its "consciousness," the IDF stuck to the belief in the notion of wars of conviction.

---

5 Neve Gordon and George A. Lopez, *Terrorism in the Arab-Israeli Conflict*, In A. Valls (ed.), *Ethics in International Affairs* (Lanham: Rowman & Littlefield, 2000), p. 106.

6 *Haaretz*, 31 October 1986.

7 Efraim Inbar, *Rabin and Israel's National Security* (Baltimore: Johns Hopkins University Press, 1999), pp. 104–5.

8 Ibid., pp. 34–5.

9 *Yediot Aharonot*, 30 January 1995.

10 Interview with Shelly Yechimovich, "Meet the Press" show, Israeli Television, Channel 2, 16 March 2002.

11 Ariel Sharon in a Speech to the Nation, Israeli Television, Channel 1, 2 April 2002.

Inspired by RMA thinking and under the impression of policing missions in the territories, IDF commanders substituted their commitment to victory for a commitment to "leverages and effects." Even the failure in the Second Lebanon War to bring Hezbollah to acknowledge its bad condition within a few days via "leverages and effects" did not trigger any second thoughts regarding their efficiency. The IDF rather concluded that the "leverages and effects" should merely be improved.[12]

During the Second Lebanon War, Prime Minister Ehud Olmert said it was "a day-to-day struggle, in which we ought to project cold bloodedness and determination."[13] This understanding was shared by an array of Israeli officials, such as Chief-of-Staff Dan Halutz,[14] Deputy Chief-of-Staff Moshe Kaplinsky,[15] Former IAF Chief General (retired) Eytan Ben Eliyahu,[16] IAF Chief Eliezer Shkedy,[17] and Minister of Transportation (former Defense Minister) Shaul Mofaz.[18] In mid-November 2006, Olmert delivered a speech on the difficulties entailed in coping with terror challenges. War against terrorism cannot be finished in one go, he said, reminding those who were demanding a Defensive Shield-like operation in Gaza that terror from the West Bank continued despite that operation.[19]

Against the backdrop of the Intifadas, and to a great extent triggered by them, a serious intellectual effort was made in the IDF to understand the nature of asymmetrical conflicts and attrition situations, an effort that is discussed in Chapter 2.

During the course of Israel's protracted asymmetrical wars since the 1950s, the IDF has learned two major lessons. First, whereas in its regular wars Israel could more easily apply an offensive strategy, although in many cases strategic offense was accompanied by defense – either at different levels-of-war (i.e., the operational or tactical), or at different fronts – the conduct of LICs required a balanced offensive/defensive approach. In LIC situations both offense and

---

12 A Report by former Chief-of-Staff Dan Shomron. Alex Fishman, "For Your Attention, Gabbi [Ashkenazi)," *Yediot Aharonot Weekend Supplement*, 26 January 2007.

13 *Yediot Aharonot*, 16 July 2006.

14 <http://www.ynet.co.il/articles/0,7340,L-3278255,00.html>.

15 <http://www.nytimes.com/2006/07/18/world/middleeast/18cnd-mideast.html?ex=1310875200&en=922e5a114f378782&ei=5088&partner=rssnyt&emc=rss>.

16 Eytan Ben-Eliyahu, "The Rear Will be the Decisive Factor," *NRG News*, 16 July 2006 <http://www.nrg.co.il/online/1/ART1/449/981.html>.

17 <http://www.ynet.co.il/articles/0,7340,L-3276824,00.html>.

18 <http://www.nrg.co.il/online/1/ART1/449/775.html>.

19 Barak Ravid, "Olmert: In This War There is No Once and For All." Available at <http://www.nrg.co.il/online/1/ART1/506/249.html>.

defense are almost equally important, each form contesting different aspects of the challenge. Offense had substantial advantages; it instilled among Palestinian guerrillas, and terrorists in particular, a strong sense of danger, thereby limiting their freedom of action and forcing them to hide. But defense was necessary in order to defend Israeli civilians and troops and to strengthen the sense of security and personal safety among civilians under terror attacks. The elusive nature of terror made defensive measures even more crucial.

As early as the 1950s, the IDF, against its will but in response to the political echelon's demands, invested in defensive measures along the borders in order to strengthen the periphery citizens' sense of security and help them withstand terror attacks. Throughout the years, Israel's defensive posture in its asymmetrical wars was based on two main elements: Buffer zones, which included security fences on the Lebanese border, along the Jordan Valley, and between Israel and the territories, equipped with electronic devices; and a system of shelters and protected spaces to protect civilians under missile attacks. In recent years Israel has also been engaged in developing such means as anti-rocket systems.

The second lesson learned by Israel during its protracted LICs was that as a Western democracy, its counterinsurgency efforts must be accompanied by a commitment to comply with high moral standards, particularly using force discriminately and proportionally and respecting civil liberties as much as allowed by the security needs and conditions. This commitment has been inspired by Jewish and liberal-democratic values, but was also affected by the acknowledgement that internal and external legitimacy was necessary. Not only has Israel stressed its commitment to fight morally, it has also educated its troops in the spirit of the "purity of arms" (On the purity of arms and the IDF's code of ethics see Chapter 4).

The more a military is exposed to LICs, the greater the chances of a reverse gap, that is, too much intellectual efforts and material resources being invested in LICs at the expense of HICs. Years of preoccupation with policing missions in the territories have weakened the IDF's operational skills against regular and semi-regular/hybrid challenges. Many of the IDF's weaknesses exposed during the Second Lebanon War derived from the nature of the IDF's activities in the territories. This should not have come as a surprise. Already in 1998, more than a decade after the outbreak of the First Intifada, Martin van Creveld warned that "ten years of trying to deal with the Intifada have sapped the IDF's strength by causing troops and commanders to adapt to the enemy. The troops now look at mostly empty-handed Palestinian men, women, and children as if they were in fact a serious military threat."[20] As if predicting the IDF's poor

20 Van Creveld, *The Sword and the Olive*, pp. 362–3.

performance during the Second Lebanon War, Van Creveld added: "Among the commanders, the great majority can barely remember when they trained for and engaged in anything more dangerous than police-type operation; in the entire IDF there is now hardly any officer left who has commanded so much as a brigade in real war."[21] Before the war, at least two General Staff members, General Yishai Beer and General Yiftah Ron-Tal, warned of the negative implications of the preoccupation with policing missions in the territories, claiming that the IDF was losing its maneuverability and capability to fight conventionally.[22]

And indeed, fighting terrorists and suicide bombers has become the IDF's sole source of combat experience. Israeli troops became used to confronting a numerically inferior opponent, who was poorly trained and equipped, while they benefitted from excellent tactical and operational intelligence provided by AMAN and the General Security Service (GSS), massive logistical and technical support, and a familiarity with the combat environment in which they had been fighting for many years. During the Second Intifada, Israel reached an 85 percent success rate of targeted killing attempts.[23] These high rates were achieved thanks to high-quality intelligence, based on a combination of SIGINT, HUMINT, and a variety of vision devices, such as unmanned aerial vehicles (UAVs), which led the Israelis to their targets, and joint inter-service (ground forces-military intelligence-IAF-police-GSS) activity that allowed the IDF to overcome the problem of targeting an elusive enemy.[24]

However, none of these advantages were available during the war in 2006. Hezbollah's fighters were highly motivated, well-trained and equipped; the tactical intelligence that was provided to the ground troops by AMAN was of low quality, if any, compared to that provided by the GSS in the territories and in South Lebanon prior to the 2000 withdrawal; commanders lacked experience in operating large formations in general and armor formations in particular; logistical support was mostly ineffective; and some of the fighting took place in an unfamiliar terrain.

In the territories, the IDF used to protect soldiers from small-arms fire by sheltering them in the houses of the local population. Based on this experience, in Lebanon soldiers were ordered to take shelter in a similar manner, ignoring the fact that Hezbollah was using sophisticated anti-tank guided

21 Ibid., p. 363.

22 Ibid., p. 131.

23 Interview with IAF Chief General Shkedy, *Janes Defence Weekly*, Vol. 42, No. 1 (5 January 2005), p. 34.

24 Ibid.

missiles (ATGMs). Consequently, in the village of Debel, out of 110 reserve soldiers residing in a single house, nine soldiers were killed and 31 wounded when Hezbollah destroyed the house using ATGMs.[25]

After having applied post-heroic warfare in the territories with great success, the IDF failed to understand that such warfare did not suit a war of higher intensity. In the territories, the IDF was much stronger, and the conditions were much more suitable for the use of technology that could save troops' lives. "When you are so strong, [...] and given the existence of other [i.e., technological] means one could employ in order to do the job, it does not make sense to risk the lives of troops. In a current security situation this may well be the right policy, but in a state of war, it becomes inappropriate," explained General (retired) Yoram Yair.[26]

Technological Developments

Technological developments have affected Israeli military thought, and have challenged traditional thinking and practice. The plethora of aspects referred to below reflects the extent to which technology has become a dominant factor.

The Ascendancy of Firepower

Force multipliers have always been a cornerstone of Israeli operational art. Whereas in the past the force multipliers were maneuver-oriented, as a result of the ascendancy of firepower they have undergone a gradual metamorphosis. Offense, first strike, indirect approach, concentration, and *Blitzkrieg* via maneuver have been replaced by force multipliers achieved via fire.

Offense

The importance of offense is twofold: first, no active battlefield victory can be achieved without offense; second, the other force multipliers are in a way derivations of offense or defense. This is why any consideration of force multipliers must put offense and defense in the center. Two criteria that can assert which of the two forms had better be pursued are, first, which form is able to achieve victory with a smaller size of force; second, which one enables fulfilling the

25 Shosh Mula and Assefa Peled, "Testimonies from the Heart," *Yediot Aharonot Weekend Supplement*, 17 November 2006; Shelah and Limor, *Captives in Lebanon*, pp. 335–6.

26 Yossi Yehoshua, "Declining Values," *Yediot Aharonot Weekend Supplement*, 13 July 2007.

missions with lesser casualties. In the theoretical literature, opinions on the relative strength of offense and defense range from the one held by defense advocates (e.g. Clausewitz), via the one that believes in a mixed-approach (e.g. Bernhardi and Liddell Hart), to the one held by offense advocates (e.g. Napoleon, Jomini, Moltke, Schlieffen, and Fuller). The latter stress the attacker's advantage that stems from his ability to choose the time and place of confrontation, which allows him to concentrate his forces to his suiting. Defense advocates, on the other hand, underline the protection the defender enjoys from his fortifications (if any), his acquaintance with the territory, and his greater determination, which stems from the fact that he often defends his home territory.

Most IDF commanders have not been aware of this spectrum of opinions as they have not been sufficiently exposed to military theory, but common sense and good intuition have compensated for their lack of theoretical knowledge, and have led IDF commanders to prefer offense, which was not only considered a force multiplier but also a means of transferring the war to the enemy's territory so as to avoid the need to absorb enemy attacks on Israeli soil. These aspects of the offense/defense dilemma were reflected in the aforementioned debate between offense and defense advocates in Israel during the early 1980s, which developed as a late reaction to the constraints on maneuver and offense during the 1973 October War. The staunchest advocates of the traditional offensive approach, such as General Israel Tal and Brigadier-General Dov Tamari, stuck to their belief in the advantages of offense, but were challenged by those who believed in a more balanced approach.[27]

The ascendancy of firepower over maneuver has not stripped offense from all of its advantages. The attacker that enjoys a technological edge – in this context, Israel – can dramatically increase the level of attrition it inflicts on its enemy through fire, particularly if it has dominance in battlefield knowledge (DBK), while ensuring greater survivability for its own troops by using fire from afar.[28] Another option for the attacker is to transfer the war to the enemy's

---

27 See note 48.

28 See Steven Metz, *Armed Conflict in the 21st Century: The Information Revolution and Post-Modern Warfare* (Carlisle, PA: Strategic Studies Institute, April 2000), pp. 31–3. The goals set out for the US military forces a decade into the twenty-first century by former secretary of defense Cohen and former chairman of the Joint Chiefs of Staff General Shalikashvili were that they possess "dominant battlefield knowledge," "full dimensional protection," "dominant maneuver," and "precision strike" ability from long distances. Michael O' Hanlon, *Technological Change and the Future of War* (Washington, DC: Brookings Institution, 2000). See also Gordon R. Sullivan and James M. Dubik, "War in the Information Age," *Military Review*, Vol. 74, No. 4 (April 1994), pp. 55–6.

territory via fire. As for the defender, the ascendancy of firepower can be an asset, too. As former Prime Minister Ehud Olmert stated, "once upon a time we were frightened by the idea that Syrian armored divisions could just role onto Israeli territory. Now we live in a new reality, having the tools for containing ground attack without capturing even one inch of Syrian land, and [...] capable of winning such [a] campaign [...] from a distance."[29]

Indirect Approach

Liddell Hart's biographer, who, like the great thinker himself, presented the IDF as one of Liddell Hart's best pupils, cited Yitzhak Rabin's admission that the implementation of the indirect approach during Israel's War of Independence was designed to overcome its inferiority in arms and numbers and the vulnerability of its people and territory.[30] Although IDF commanders were hardly exposed to Liddell Hart's writings,[31] they have applied the indirect approach based on common sense and good intuition.

This was true for the 1940s, 1950s, and the 1960s. In the 1970s, however, only little was left of the traditional Israeli indirect approach, which had been the jewel in the crown of its operational art. Its spirit, however, was sporadically preserved and could occasionally be found, particularly among the older generation of IDF commanders, such as in the following military operations: the crossing of the Suez Canal and the encircling of the Egyptian 3rd Army in October 1973, Ariel Sharon's plan to envelop Syrian troops in Lebanon during the 1982 First Lebanon War, or the battles waged by the paratrooper brigade under the command of Colonel Yoram Yair during that war. Another example is from the 2006 Second Lebanon War: Before the war, the IDF had planned an operation based on a "sophisticated blend of amphibious, airborne and ground penetrations to swiftly extend deep into the front, before rolling back, so as to destroy Hezbollah's positions one by one from the rear, all the way back to the Israeli border."[32] But the plan was eventually not implemented. Israeli ground activity consisted of transparent maneuvers, something that operational art and the indirect approach cannot tolerate, unless such maneuvers are carried out for deception purposes. Had the IDF really been committed to its sophisticated tradition, its ground operations would have opened by quickly

29 Nahum Barnea and Shimon Shiffer, interview with Prime Minister Ehud Olmert, *Yediot Aharonot New Year Supplement*, 29 September 2008.

30 Brian Bond, *Liddell Hart: A Study of His Military Thought* (London: Cassell, 1977), p. 246.

31 Tuvia Ben-Moshe, "Liddell Hart and the Israel Defense Forces: A Reappraisal," *Journal of Contemporary History*, Vol. 16, No. 2 (April 1981), pp. 369–91.

32 Edward N. Luttwak, "Misreading the Lebanon War," *Jerusalem Post*, 21 August 2006.

outflanking and encircling the enemy and using the element of surprise for capturing the northern parts of Southern Lebanon first. An indirect approach à la Sun Tzu or Liddell Hart would have caused confusion among the enemy ranks and would most probably have brought about its psychological collapse much better than the Clausewitzian direct approach, which helped Hezbollah recover and stand strong.

There is no doubt that with the decline of maneuver, the indirect approach on the ground has lost weight and has not been applied as frequently as before, but as pointed out above, some experts believe that sophisticated firepower has substituted for Liddell Harts' traditional indirect approach and could be considered a more modern application of it.[33]

Concentration of Forces

The assumption that Israel might find itself outnumbered when fighting against an Arab war coalition dictated operating in accordance with the "logic of the few," which required determining in advance where the main effort would be concentrated in order to amass sufficient forces at that point so as to compensate for the IDF's general quantitative inferiority. Concentration of forces on the ground took time, and so was diverting forces from one front to another. With the ascendancy of firepower, concentration of fire has become more important than concentration of forces, and the availability of long range and precise firepower has made it possible to concentrate or disperse fire within a very short time. There are at least two problems in the new reality, though: first, in counterinsurgency operations, a concentration of fire has a much smaller effect than ground maneuvers; second, players that proved to be poor in maneuvering have taken advantage of the ascendancy of firepower to acquire rockets and missiles that would enabled them to balance, at least to some extent, their military inferiority and to concentrate or disperse fire. These are lessons that the IDF learned in recent decades.

RMA-inspired concepts that have been adopted by the IDF since the 1990s have even further distanced it from the old and much simpler notions of concentration of forces and center of gravity. For example, the concept of "diffused warfare," which was crystalized by Haim Assa and Yedidya Yaari. The concept was based on the assumption that, first, campaigns consisting of horizontal clashes between rival forces, which entail breaking through the opponent's layers of defense and proceeding along distinct lines, with distinct start and finish lines, came to an end. Second, that they were substituted by diffused

33 Shmuel Gordon, *The Bow of Paris* (Tel Aviv: Poalim, 1997) [Hebrew], particularly pp. 320–2.

confrontations that were supposed to takes place simultaneously over the entire "battlespace," distributing the force's mass to a multitude of separate pressure points, rather than concentrating it on assumed centers of gravity.[34] Diffused warfare, which constituted a challenge to the notion of concentration, proved to be ineffective during the 2006 Second Lebanon War. It took Israeli strategy back to the times when achievements on the battlefield had to be accumulated in a bottom-up process, and numerous tactical successes had to be translated into operational or strategic success.

First Strike

Although delivering a first strike has never been an explicit principle in Israeli war doctrine, Israelis have always believed that striking first, if only possible, preferably with the consent or backing of a great power or a superpower, could increase the chances of demoralizing the enemy, and achieving battlefield success. And, indeed, both in 1956 and 1967, the first strike did have a considerable negative effect on the Arab war effort.

Whereas in the past the time dimension constituted a crucial factor in planning a first strike because of the need to mobilize and deploy sufficient ground forces to that end, which also exposed the intention to attack, the considerably enhanced firepower capabilities have made it possible to prepare a first strike without exposing this intention and landing the first strike within minutes. The strengthening of firepower capabilities has also enabled a near-real-time retaliation in cases of a surprise attack or political constraints that make the launching of a first strike impossible.

Blitzkrieg

Blitzkrieg was one of Israel's typical expressions of operational art. Both in 1956 and 1967 it brought about the enemy's psychological collapse, shortened the war's duration, and achieved a decisive victory on the battlefield before the superpowers or a significant number of Arab expeditionary forces had the chance to intervene. It did so due to the fast advance of Israeli forces in enemy territory, which was effectively supported by a decentralized logistical system and air supremacy, and enabled also through a mission command and high quality field commanders.

The shift from maneuver-enabled battlefield to a firepower-dominated one was detrimental to the traditional *Blitzkrieg*, because the latter needed maneuver in order to be successfully carried out. Effective anti-tank and anti-aircraft weapons that were introduced in the battlefield on the Arab side, alongside the

---

34 Haim Assa and Yedidya Yaari. *Diffused Warfare* (Tel Aviv: Yediot Aharonot, 2005) [Hebrew].

saturation of the battlefield with troops since the 1970s, put obstacles on the IDF's way both in the skies and on land and made *Blitzkrieg* much less feasible, even almost obsolete.

The ascendancy of firepower was reflected during the 1969/70 War of Attrition and the 1973 October War, where the missile had the upper hand over the aircraft. Many Israeli aircraft were downed – 14 during the War of Attrition, and 102 aircraft and 5 helicopters during the October War.[35] In 1973, the saturation of the battlefield with both forces and fire as well as the constraints on mobility as a result of the nature of the terrain, particularly the sandy Sinai Desert, have imposed great difficulty for air offensives and ground maneuvers.

These constraints, however, have been accompanied by at least two opportunities proceeding from the ascendancy of firepower. First, offense could benefit from improved firepower, too. For example, on June 9, 1982, thanks to advanced technological systems, the IAF, almost single-handedly, successfully destroyed the Soviet-made Syrian SAM network in the Bekah Valley, proving that the aircraft was again superior to the SA missile, and that offense can again have the upper hand. Second, the ascendancy of firepower has made it possible to transfer the war to the enemy's territory via firepower. These benefits, however, seems to apply to HICs rather than LICs, as was proved in 2006, when the IDF's firepower superiority was not translated into battlefield success.

An Israeli firepower-based approach was first introduced in the mid-1980s, when Colonel (retired) Prof. Amnon Yogev, chief of military R&D, offered a new security model for Israel. Yogev pointed to the obsolescence of the maneuver-oriented big platforms-based, quick, decisive victory model, which, in his opinion, had reached its culminating point in 1967, and since then had become less effective and more costly. He claimed that an alternative, high-tech-based armament model would cost half the budget despite the high cost of each system; that it would be much more effective, thanks to the precision, long range and destructiveness of the armament; and that it would release Israel from its heavy dependence on early warning. To Yogev's credit it must be stressed that as much as he was committed to technology he understood its limits, such as the inability of airpower to achieve decisive victory single-handedly.[36]

As was described in Chapter 2, in April 2006, the IDF issued a new operational doctrine, which was based on firepower capabilities. The belief that technology could win wars, however, was shattered during the 2006 Second

35 Benny Morris, *Righteous Victims* (Tel Aviv: Am Oved, 2003) [Hebrew], pp. 342, 406.

36 Amnon Yogev, *A Future Defense Model for Israel* (Tel Aviv: Jaffee Center for Strategic Studies, 1986); Amnon Yogev, "Israel's Security in the 1990s and Thereafter," Alpayim 1 (1989), pp. 166–85.

Lebanon War that broke out a few months later. The process in which the non-material aspects of Israeli strategy and tactics were overshadowed by technology was something military experts in the US had warned from before the war, arguing that overreliance on firepower for battlefield success might have a weakening effect on traditional military capabilities such as close combat or combat intelligence.[37]

In recent decades, nonstate actors have been playing in the technological playground, acquiring capabilities that in the past would have been found only in the hands of state actors. With firepower capabilities at their disposal – simple and relatively cheap rockets of various ranges, such as Qassam, Grad, and Fajr – Palestinians and Hezbollah have acquired the ability to attack Israel's rear, demoralize its civilians and disrupt social and economic activities in northern or southern Israel. Their stockpile of missiles and rockets has been sufficient for the daily launching of dozens to hundreds of rockets along weeks of fighting. In late-2013 Hezbollah possessed an arsenal of over 80,000 missiles and rockets of all calibers. It has also tried to acquire state-of-the art surface-to-air batteries and surface-to-sea systems, and has been developing drones.[38]

However, despite the fact that the availability of enhanced firepower has somewhat blurred the power gap between the capabilities of Israel and its nonstate enemies, this gap has remained considerably wide, tempting Israel to stick to the heavy reliance on firepower rather than maneuver. Although not admitted explicitly, Israeli commitment to the use of firepower instead of maneuver has only increased, and was manifested in the relatively recent LICs – the 2008/9 Operation Cast Lead, the 2012 Operation Pillars of Defense, and the 2014 Operation Protective Edge against Hamas. In all of these operations the IDF refrained from capturing the entire Gaza Strip, let alone the toppling of the Hamas regime, because this might have meant maintaining a military presence in the Gaza Strip for an unknown period, likely to entail a significant number of casualties.[39]

---

37 See, for example, Stephen Biddle, *Military Power: Explaining Victory and Defeat in Modern Battle* (Princeton: Princeton UP, 2004); Stephen Biddle, "Military Power: A Reply," *Journal of Strategic Studies*, Vol. 28, No. 33 (June 2005), pp. 453–69.

38 *Haaretz,* 23 May 2012 <http://www.haaretz.com/news/diplomacy-defense/idf-israel-in-range-of-nearly-65–000-hezbollah-iran-syria-rockets.premium-1.432012>; Yaakov Lappin, "Israel vs. the Iran-Hizballah Axis," *BESA Perspectives*, No. 221 (14 November 2013) <http://besacenter.org/perspectives-papers/israel-vs-iran-hizballah-axis/>.

39 Amos Harel, "The Hundred Years War," *Alaxon*, 14 April 2013. <http://alaxon.co.il/article/%D7%9E%D7%9C%D7%97%D7%9E%D7%AA-%D7%9E%D7%90%D7%94-%D7%94%D7%A9%D7%A0%D7%99%D7%9D-%D7%A2%D7%9C-

Success in decreasing the number of casualties on both sides using sophisticated active-defense systems (which are part of Israeli firepower capabilities) has made it more difficult for Israel to justify large-scale punitive operations. Such systems merely frustrate the other side rather than inflict a heavy toll on it, and as such they might be of little deterrent value, if at all.[40] This problem has not yet been fully internalized by Israel's military doctrine heads. To cite former director of the Research and Development (R&D) directorate at the Ministry of Defense, General Isaac Ben-Israel, "Here in Israel we have realized that our relative advantage in frustrating terrorist attempts is carried out mainly through technology. Our success rate is very high."[41]

The Notion of "Small But Smart Military"

Another fact tended to be ignored by Israeli military thinking is that coping with LIC challenges usually requires more troops than high-intensity conflicts, as a small number of troops using high-tech equipment would be insufficient for destroying a sophisticated guerrilla force, capturing the terrain from which guerilla warfare is conducted, achieving decisive victory on the battlefield, or destroying rockets launchers used by insurgents against populated areas.

The Second Lebanon War proved that it is difficult, perhaps impossible, to destroy a sophisticated guerrilla force via RMA warfare. As General (retired) Amiran Levin pointed out, the overreliance on precision technology was one of the major reasons for the IDF's malfunctioning in the war.[42]

The notion of "small but smart" military was affected by the better effectiveness of weapon systems, their rising cost, and the need to decrease the number of weapon systems and troops on the battlefield in order to reduce their vulnerability. It was coined by Chief-of-Staff Dan Shomron in the late 1980s, and adopted by subsequent chiefs-of-staff like Moshe Yaalon and Benny Gantz.

On the other hand, it is true that the more force one uses during asymmetrical wars, the greater the chances of collateral damage, mistakes, or accusations of excessive use of force.

The implications for the IDF's reserves have been clear. After having relied on reserves for many decades (the reserves comprised some 3/4 of the IDF's

---

%D7%A9%D7%9B%D7%95%D7%9C-%D7%95%D7%A0%D7%A4%D7%92%D7%A2%D7%99%D7%9D/>.

40 Avi Kober, "Iron Dome: Has the Euphoria Been Justified?" *BESA Perspectives*, No. 199, 25 February 2013; Uzi Rubin, "Iron Dome: A Dress Rehearsal for War?" *BESA Perspectives*, No. 173 (July 2012); "Iron Dome in Action: A Preliminary Evaluation," *BESA Perspectives*, No. 151 (24 October 2011).

41 <http://www.samliquidation.com/chabad_9d.html>.

42 Amos Harel, "A Flawed Operational Conception," *Haaretz*, 10 December 2006.

ground forces) in recent decades the IDF's reserves units have shrunk significantly. Instead, the IDF adopted the notion of a "small but smart" military. This, however, has been incompatible with the empirically based paradox that LICs require big armies,[43] which requires maintaining a considerable number of reserve units, which in turn raises the challenge of insuring a reasonable level of professionalism among the reserves.

This was demonstrated in the 2006 Second Lebanon War, during which the IDF suffered from a shortage in ground troops as a result of years of insufficient investment in the reserve troops. Prior to the outbreak of the war the IDF failed to understand that a significant number of reserve units are needed in LICs, too. "Conventional war is no longer our top priority," explained General Danny Van Buren, chief of the IDF's reserve forces, two months before the outbreak of the war.[44] Based on this premise, the IDF phased out some reserve units, reduced the number of reservists activated, cut nearly in half the days served per year from 30 to 14, activated reserve units only for training (not combat or guard duty), and lowered the maximum age for reservists from 46 to 40. In the war the army paid dearly for these changes.

Victory from the Air

Another aspect of the technological surge of recent decades has been the belief in the airpower's ability to achieve decisive victory in war thanks to aircraft's superior firepower, higher maneuverability and greater flexibility in comparison with ground forces.

For many years Israeli military doctrine considered the ground forces the backbone of any large-scale military operation, both in HICs and asymmetrical war against nonstate players with irregular or semi-regular forces. Air power was considered a major factor in creating the necessary conditions for battlefield success, but was never considered a factor that could decide a war by itself. This has changed. Although former Chief of Ground Forces, General Benny Gantz, denied that anyone in Israeli military leadership had really ever held the view that air power alone could decide a war,[45] the belief was not merely in the back of their minds. In a discussion at the National Defense

---

43 Avi Kober, "Operational and Technological Incentives and Disincentives for Force Transformation," in Stuart Cohen (ed.), *The New Citizen Armies* (London: Routledge, 2009), pp. 77–91; Avi Kober, "Can the IDF Afford a Small Army? *BESA Perspectives*," No. 209 (18 July 2013).

44 <http://www.usatoday.com/news/world/2006-06-05-israel-army_x.htm>.

45 Barbara Opall-Rome, Interview with Major-General Benjamin Gantz, Chief of the IDF's Army, *Defense News*, 28 August 2006, p. 38.

College in January 2001 IAF Chief Dan Halutz argued that "many air operations were generally implemented without a land force. [...]. This obliges us to part with a number of anachronistic assumptions. First of all, that victory equals territory. Victory means achieving the strategic goal and not necessarily territory. I maintain that we also have to part with the concept of a land battle."[46] In 2002, still as IAF Chief, Halutz referred to the IAF's capabilities: "Air power alone can decide, [... or at least] be the senior partner to such decision."[47] As a result of the priority given to airpower, ground forces budgets were cut, one of the results being Israeli tanks lacking active protection systems, smoke obscuration equipment, etc.[48]

In his testimony before the Winograd Commission after the Second Lebanon War, Halutz reiterated his belief that given the ascendancy of firepower on the battlefield, the air force, thanks to its outstanding fire capabilities, could play a dominant role on the modern battlefield.[49] He was so confident that airpower could do the job alone, or almost alone, that during the 2006 Second Lebanon War he did not provide the government with any real alternative plan until the latest stage of the war. The operational order issued by the General Staff on July 13, 2006, at the outbreak of the confrontation, which at this stage referred to a campaign (codenamed "Just Reward"), not a war, described the upcoming operation as a stand-off fire-based protracted offensive.[50] According to Defense Minister Amir Peretz, Chief-of-Staff Halutz believed in obtaining decisive victory via massive firepower, and did not change his views until the end of the war.[51] Halutz did not deny this, yet admitted that not preparing for a ground operation had been a mistake, affected by his failure to foresee the lengthy duration of the operation (33 days).[52] General (retired) Yossi Peled criticized the new state of mind that had permeated the IDF: "Something very bad has happened to the IDF in recent years. We have lost the balance between the forces within the IDF, giving credit to the IAF's ability to solve any problem.

---

46 Zeev Schiff, "The Foresight Saga," <http://www.haaretz.com/hasen/spages/749268.html>.

47 Amnon Lord, "The Air Went Out," *Makor Rishon*, 2 November 2006 <http://www.makor-rishon.co.il/show.asp?id=14091>.

48 <http://www.ynet.co.il/articles/1,7340,L-3297389,00.html>.

49 Halutz's testimony before the Winograd Commission <http://www.vaadatwino.org.il/pdf/חלוץ%20דן%20תמליל.pdf>, p. 16.

50 Alex Fishman, "The Five-Day War," *Yediot Aharonot Weekend Supplement*, 17 April 2007.

51 Alex Fishman, "Thanks to The Censorship," *Yediot Aharonot Weekend Supplement*, 11 May 2007.

52 Ibid.

A golden calf was created and named technology; many believed it could win the war."[53]

The inability to achieve victory from the air should not have come as a surprise. In February 2004, former Chief-of-Staff Moshe Yaalon briefed the Knesset Foreign Affairs and Defense Committee on the rocket threat from Lebanon, stressing that diminishing the rocket fire without operating on the ground might take weeks.[54] Former IAF Chief and former Deputy Chief-of-Staff General David Ivry, too, disagreed with Halutz, arguing that airpower could not win the war against terrorism alone.[55]

Before the outbreak of the Second Lebanon War, Chief of Northern Command General Udi Adam and his deputy General Eyal Ben-Reuven, had prepared a large-scale ground operation (codenamed "Mey Marom"), but the highest political and military echelons were against its implementation.

On the second night of the war, after the IAF's success in destroying Hezbollah's long-range and middle-range launchers, Chief-of-Staff Halutz informed Prime Minister Ehud Olmert that the IDF had won the war.[56] Less than a week after the war started Halutz informed the Knesset Foreign Affairs and Defense Committee that the IDF was achieving its goals by employing airpower and artillery without launching a ground operation into Lebanon.[57] As the war progressed, however, it became evident that there was a great disproportion between the unprecedented number of combat sorties carried out by the IAF – 11,897, more than the number of sorties during the 1973 October War (11,223), and during the 1982 First Lebanon War (6,052)[58] – and their impact. The impact on Hezbollah's capability to continue fighting and launching on a daily basis hundreds of Katyusha rockets onto Israeli territory proved to be limited, despite improved hunting tactics applied by the IAF.

On June 28, 2006, Chief of AMAN General Amos Yadlin, who, like Halutz, also hailed from the air force and had rejected the idea of ground operations, admitted that it was necessary that the war ended with an Israeli victory, and that to that effect there was an urgent need to operate on the ground. The IDF, which had defeated the Arab states within six days, could finish the job quickly by sending in two divisions, Yadlin said.[59]

---

53 Ari Shavit, Interview with General (retired) Yossi Peled, *Haaretz Weekend Supplement*, 20 October 2006.

54 *Yediot Aharonot*, 22 January 2007.

55 Schiff, "The Foresight Saga."

56 *YNET*, 27 August 2006 <http://www.ynetnews.com/articles/0,7340,L-3296031,00.html>.

57 17 July 2006 <http://www.isracast.com/index.aspx?y=2006&m=7>.

58 Isaac Ben-Israel, "The Missile War," an unpublished paper.

59 *Haaretz*, 23 January 2007; Shelah and Limor, *Captives in Lebanon*, pp. 212–13.

Amazingly, despite the criticism raised over the overreliance on technology that had been demonstrated during the Second Lebanon War, some airpower experts, like Shmuel Gordon, still believed that the combination of high-quality intelligence, airpower and Special Forces was the ultimate counter-terror formula for winning the war against Hezbollah.[60] In 2014 IAF Chief Amir Eshel and other IAF senior officers still believed that with excellent intelligence and the arms and aircraft available at the IAF's disposal, it could win the war on its own, without the IDF having to maneuver its way into enemy territory.[61]

If it is true that the IAF has eroded the role and status of the ground forces, it is all the more so when it comes to the Israel Naval Force (INF). The INF is a small arm, with a limited number of vessels, and a relatively small budget. Some of the INF's tasks only complement those of the other arms, which explains why it has lagged behind, and been overshadowed by, the other arms. Its more natural tasks are ensuring Israel's sovereignty at sea and free navigation from and to Israel, and protecting its shores by keeping continuous presence at sea and thwarting hostile naval threats. Other missions are assisting ground operations in various ways, e.g., landing troops from the sea, applying fire from the sea, and outflanking the enemy from the sea. Reflective of the INF's limited status is the fact that the IDF's doctrines have focused on AirLand battle, which consists of cooperation between air power and ground troops.

In recent decades, however, the INF's importance has been on the rise after having become the third leg in Israel's strategic triad, which consists of strategic air force, surface-to-surface missiles, and advanced submarines with strategic capabilities, a triad that is supposed to ensure Israeli second strike capability.

Network-Centric Warfare

Fascinated by the quick developments in the area of information technology, the IDF has enthusiastically followed in the footsteps of the American army in this respect, too. One of the expressions of this path has been the belief in network-centric capabilities. These capabilities are meant to translate information superiority into increased shared awareness, speed of command, operational tempo, lethality, survivability, and synchronization, in the spirit of John Boyd's Observation-Orientation-Decision-Action loop. This loop means

60 See for example, Shmuel Gordon, "Winning from the Air is Still Possible," <http://www.ynet.co.il/articles/1,7340,L-3288914,00.html>.

61 Ron Ben-Yishai, "Israel Air Force is deadlier than ever," *YNET*, 8 May 2014 <http://www.ynetnews.com/articles/0,7340,L-4517002,00.html>.

that "the opponent must be observed to gather information; the attacker must orient himself to the situation or context; then decide and act accordingly."[62]

In the mid-1990s, the IDF launched its own version of Network-Centric Warfare (NCW), which gained the name "Ground Forces Digitalization" program (TZAYAD, in Hebrew).[63] During the Second Lebanon War, however, the expectation that a technological superiority-based accurate intelligence would be available proved to be exaggerated. Due to gaps in intelligence, outdated intelligence, or failure to distribute intelligence to the troops on the battlefield in real time, or near real time, IDF troops often operated blindly, occasionally being surprised by an inferior enemy,[64] and doubts were raised regarding the expected cost-effectiveness of NCW.[65]

Command and Control over Plasma Screens

Another rotten fruit of the over-reliance on technology has been the tendency of senior IDF commanders to run the battle from their headquarters, instead of at the head of their troops on the battlefield, as happened during the Second Lebanon War.[66] Following the war, both Chief-of-Staff Dan Halutz and former Deputy Chief-of-Staff Matan Vilnai pointed to this phenomenon as one of the reasons for the difficulties faced by the IDF on the battlefield. Conducting the war over plasma screens may have been compatible with the assumption that on a battlefield where enemy forces are destroyed by stand-off precision fire, optimal command and control is achieved from control centers. At the same time, however, "it may have changed the focus of our command," Halutz admitted.[67] Vilnai's diagnosis was that what the IDF had lacked in that war was a

62 ores David S. Alberts, John J. Garstka, and Frederick P. Stein, "Network-Centric Warfare: Developing and Leveraging Information Superiority," <http://www.dtic.mil/cgi-bin/GetTRDoc?Location=U2&doc=GetTRDoc.pdf&AD=ADA406255>; Carlo Kopp, "Understanding Network Centric Warfare," Australian Aviation (January/February 2005) <http://www.ausairpower.net/TE-NCW-JanFeb-05.html>.

63 <http://www.defense-update.com/products/z/zayad.html>.

64 Ibid., p. 38. See also Shelah and Limor, *Captives in Lebanon*, p. 288.

65 Anthony H. Cordesman, *Preliminary "Lessons" of the Israeli-Hezbollah War* (Washington DC: Center for Strategic and International Studies, 11 September 2006), p. 15 <http://www.csis.org/media/csis/pubs/060911_isr_hez_lessons.pdf>.

66 Shelah and Limor, *Captives in Lebanon*, p. 385.

67 <http://www.haaretz.co.il/hasite/pages/ShArtsR.jhtml?itemNo=755196&objNo=59745&returnParam=Y>.

simple command system. You can run Macdonald's using plasma screens, not a battle, he said.[68]

Cyberspace and Cyber Warfare

Unlike the four previous domains known to man – land, sea, air and outer space – the relatively new, fifth space, is man-maid. Cyberspace and the subsequent cyber warfare have developed so rapidly in recent years that their theoretical and doctrinal aspects seem to have lagged behind the new reality, similar to what had happened when the nuclear weapon had been introduced. At present, it is not yet clear if cyber warfare stands for a real revolution in the nature and conduct of war or rather constitutes just another technological means that has been employed alongside the more traditional ones. The emergence of previous spaces in which man operated had not challenged the basics of war and strategy, so that one cannot be sure whether the latest addition will challenge them either. For example, the potential of achieving political and strategic goals without resorting to armed conflict is not unique to cyber warfare, as deterrence, for example, falls into this category, too. It is true that in cyber warfare relatively weak players can enjoy disproportionate power, but this applies to the traditional challenge stemming from insurgents, too. It is also true that in cyber warfare players can enjoy anonymity and relative impunity, but had this not also been the case with terror groups long before cyber warfare was introduced? In cyber warfare the boundaries between the military and civilians are blurred, but had this not been typical of previous phenomena such as of narco-terrorism? [69] Furthermore, for cyber warfare to be considered a revolutionary phenomenon it must be proved, for example, that by using it one can achieve decisive victory by itself, which does not seem likely in the foreseeable future, if at all. Finally, even after the introduction of cyber warfare there will still be offense and defense, indirect approach, centers of gravity, first and second strike, etc., though applied in a different manner.

A lot has been written, said, and done in recent years in Israel about cyber warfare. For example: Articles and symposiums dedicated to it by the Military and Strategic Affairs Program at INSS; the establishment of a Cyber Warfare

---

68 Amira Lam, "We Betrayed our Constituency," *Yediot Aharonot Weekend Supplement*, 1 September 2006.

69 Paul Cornish, David Livingstone, Dave Clemente and Claire Yorke, *Security and the UK's Critical National Infrastructure* (London: Chatham House, September 2011), pp. vii-viii and, 1–38. See also Ian A McGhie, "Cyber-Warfare: Vital Ground, 'Emperor's New Clothes,' or strategic paralysis?" dissertation published by the Royal College of Defense Studies, July 2012 <http://www.da.mod.uk/colleges/rcds/publications/seaford-house-papers/2012-seaford-house-papers/shp-2012-mcghie.pdf/view>.

Administration in Israel in charge of coordinating the efforts of security agencies and the Israeli defense industry in developing advanced systems to deal with cyber warfare; a team headed by General (retired) Professor Isaac Ben-Israel, former director of the Research and Development Directorate at the Ministry of Defense, whose mandate was to submit to the Israeli government recommendations on how to prepare for the threat of cyber-attacks; an annual international cyber security conference at Tel Aviv University; the establishment of a cyber-staff within Unit 8200 of AMAN (the Israeli NSA), which includes representatives from the intelligence community and the IDF's Computerization Division;[70] and a new GSS unit in charge of defense against cyber-attacks. These activities have all been reflective of the curiosity and interest in the security establishment in cyber warfare. Isaac Ben-Israel's cyber warfare team recommended, among other things, that a national consultant on cyber threats be appointed and that a national headquarters for formulating a policy and monitoring its implementation be established alongside research centers in the academia, with the aim of increasing cooperation between the government, the academia and the industry.

Officials within the Israeli security establishment have considered cyber warfare a major change in the nature of war. Former Chief of AMAN, General (retired) Amos Yadlin, compared the impact of the introduction of cyber warfare to the advent of airpower in the 20th century, pointing out that it is not merely a weapon in the hands of powerful states but also one in the hands of small states, organizations and individuals.[71] In 2009, IDF Chief-of-Staff Gabi Ashkenazi defined cyberspace as a strategic and operative domain of warfare. Although Israeli officials like former Defense Minister Ehud Barak and former Chief of AMAN Amos Yadlin have stressed cyber warfare's defensive value, it can also be applied in an offensive manner, as happened in the case of the destruction of the Syrian nuclear site in 2007, or when cyber warfare was allegedly used against the Iranian nuclear program.[72]

As a result of these efforts and activities Israel has become one of the leading countries in cyber warfare. It has developed capability to act virtually from ground, naval, airborne or unmanned platforms, to act against both civil and military targets, to defend and to attack. It seems that as a result of the

---

70 Shmuel Even and David Siman-Tov, *Cyber Warfare: Concepts and Strategic Trends* (Tel Aviv: INSS, 2012) [Hebrew], p. 79.

71 Reuters and Anshek Pfeffer, "How Cyber warfare Has Made MI a Combat Arm of the IDF," *Haaretz*, 16 December 2009. <http://www.haaretz.com/print-edition/news/how-cyber-warfare-has-made-mi-a-combat-arm-of-the-idf-1.2051>.

72 <http://www.nytimes.com/2010/09/27/technology/27virus.html?hp&_r=0>.

importance attributed to cyber warfare in Israel it has become an integral part and symptom of Israel's cult of technology.

From the importance attributed in Israel to cyber warfare one can learn that despite the uncertainties it entailed, security establishment officials in Israeli have related to cyber warfare as a major change to the nature and conduct of war, and as an integral part of Israel's contemporary strategic thought.

Legal and Moral Constraints Created by International Law and Norms

A third systemic factor that has gained tremendous impact on Israeli military thought, particularly in LIC contexts, has been worldwide moral and legal norms, which Israel has been committed to abide by, not only because of its values as a Western and Jewish state, but also due to international constraints. Although legal and moral considerations have never been formally or explicitly considered as part of Israeli military thought, or recognized as such, in the post-Cold War era ethical dilemmas have only intensified and sharpened due to two main developments: the prevalence of LICs, and the destructive and murderous nature of terror. Defining terror, and establishing which countermeans were permitted and which prohibited, as well as the challenge of confronting the enemy not only on the battlefield but also in court – have been one of the greatest challenges Israeli legal bodies have faced in recent years.

This factor has made an impact in regard to two main categories: *jus ad bellum*, which defines the legal and legitimate reasons for a player to engage in war, particularly the inherent right of self-defense; and *jus in bello*, i.e., principles and laws to which a player already involved in war is expected to adhere.

Jus ad Bellum: *Preventive War as Opposed to Preemptive Attack*

A major dilemma confronting Israelis when war was becoming imminent concerned the legality and legitimacy of a first strike. The preference for first strike in Israeli military thinking might have clashed with the imperative of conducting a just war, in case Israel failed in convincing the world that the war was initiated as an act of self-defense.[73] Israeli political and military echelons have

---

73 Customary international law, articulated in 1837 by US Secretary of State Daniel Webster, which constituted the terms of reference for Just War until the introduction of a competing definition of self-defense by Article 51 of the UN Charter, defined the conditions under which war would be considered just: there is a real threat, the response is essential and proportional, and all peaceful means of resolving the dispute have been exhausted.

internalized the difference between two types of first strike – preventive war/strike and preemptive attack/strike. The first type entails the difficulty to garner international legitimacy because it is carried out when the threat is still imminent and has not yet been realized, whereas the second type refers to an actual and visible threat (e.g., enemy forces already deployed).

During the early 1950s, Israel still felt too weak to adopt an offensive approach in general and first strike in particular, and stuck to a defensive posture. Initiation of war by Israel seemed out of the question. This approach was led by Chief-of-Staff Yigael Yadin and his deputy Mordechai Maklef. In 1954, Prime Minister and Foreign Minister Moshe Sharet expressed his objection to a preventive war,[74] which he interpreted as defying the international system.[75] Even Prime Minister and Defense Minister David Ben-Gurion, who held more hawkish views, spoke against a preventive war.[76] In mid-December 1955, only ten months before Israel initiated a preventive war against Egypt, he stressed that although there was a strong inclination towards an offensive initiative, Israel should refrain from landing it, for three main reasons: First, the destruction inflicted by war; second, the danger that a third party (e.g., Britain, who had a defense pact with Egypt and Jordan) would intervene on the Arab side; and third, the fear that an Israeli offensive would bring about an embargo on arms sales to Israel that would be detrimental to the military balance with its Arab enemies.[77]

Once the decision was made to attack Egypt, and in order to bridge the gap between the need to launch a first strike for security reasons and the international legal and moral constraints entailed in it, an effort was made by Israel to disguise the first strike's preventive nature. And indeed, in justifying the initiation of the 1956 Sinai War, Ben-Gurion related to it as a preemptive act, explaining that it was a necessary reaction not only to the Egyptian-Czech arms deal but also to the formation of an Egyptian-Syrian-Jordanian alliance against Israel, the aim of which was to put an end to its existence.[78]

Article 51 authorizes self-defense only if armed attack occurs. A.N. Guiora, "Targeted Killing as Active Self-Defense," *Case Western Reserve Journal of International Law* 36 (2004), pp. 319–34. This is also compatible with the Jewish notion of obligatory war (*milhemet mitzvah*), i.e., war of self-defense, which is justified according to *jus ad bellum* criteria.

74 Jehuda Wallach, "Trends in Israeli Defense Doctrine," *Skira Hodshit* 3–4 (1987), pp. 24–9.

75 Sharet's speech on 22 May 1954 before MAPAI party members. The Labor Party archives, Beit Berl.

76 Ben-Gurion, *Yichud Ve-Ye'ud*, pp. 218–25.

77 Mordechai Bar-On, "The Sinai Campaign: Reasons and Achievements," *Skira Hodshit*, 10–11 (1986), pp. 8–9.

78 Ben-Gurion, *Yichud Ve-Ye'ud*, pp. 285–6.

Years later, Yigal Allon – one of Israel's leading strategic thinkers – explained in his *Curtain of Sand* and *Communicating Vessels* why he was opposed to a preventive war. First, Israel should postpone the outbreak of war as long as it was able to protect itself without war, for moral reasons; and second, preventive war would be difficult to defend diplomatically and might cause Israel to lose friends and bring about the imposition of an embargo of arms upon it by the great powers.[79] Allon offered the notion of a "preemptive counter-attack" as a preferred option, stressing the defensive nature of such an initiative. The so-called preemptive counter-attack was later on applied by Israel in 1967.

Israel's interpretation of the first strike as an act of self-defense was supported by one of the leading figures of the philosophy of just and unjust war, Michael Walzer. According to Walzer, preemptive strikes are justified if three conditions are fulfilled: an obvious intention on the part of the enemy to do injury; active preparations on the enemy's side that turn such intention into a positive danger; and a situation in which the risk of defeat will be greatly increased if a strike is delayed.[80] At the same time, Walzer pointed to the problem entailed in preventive strikes, in cases where an uncertainty exists with regard to the fulfillment of these conditions.[81] In 1956, Israel coped with the problem by joining a coalition with two great powers, France and Britain, who undertook to thwart any anti-Israel resolution in the UN. Years later, Israel faced such difficulty again, in June 1981, and in September 2007, when it attacked the nuclear reactors of Iraq and Syria, respectively.

Jus in Bello

After the 1953 Qibya Operation in which 69 innocent Arab civilians were killed, Chief-of-Staff Moshe Dayan realized that not only the citizens of the State of Israel but also world Jewry expected the IDF to abide by the "purity of arms" principle.[82] And indeed, diplomatic pressures were not late to come. On November 24, 1953, the UN Security Council condemned the killing of innocent civilians; Britain sent a tacit threat that it might consider standing by Jordan as required by its defense pact with the monarchy; and the US suspended economic assistance to Israel.

In the wake of Qibya, Israeli policy changed, and no longer has the killing of civilians been allowed or tolerated. Israel's counterinsurgency policy and

79 Yigal Allon, *A Curtain of Sand* (Tel Aviv: Hakibbutz Hameuhad, 1968) [Hebrew]; Yigal Allon, *Communicating Vessels* (Tel Aviv: Hakibbutz Hameuhad, 1980)[Hebrew], p. 105.

80 Michael Walzer, *Just and Unjust Wars* (New York: Basic Books, 1977), pp. 80–1.

81 Ibid., pp. 76–8.

82 Benny Morris, *Israel's Border Wars 1949–1956* (Tel-Aviv: Am Oved, 1996) [Hebrew], p. 291.

activity have been crystallized and implemented under international criticism, stemming both from its Arab enemies and international civil rights organizations like Human Rights Watch or Amnesty International. Criticism focused on the immorality or illegality of the allegedly indiscriminate and excessive use of force by Israel, and on Israeli continuous violation of the basic rights of the Palestinian population in the territories. [83]

Evidence of Israeli awareness of the need to integrate international moral and legal considerations into operational planning has been at least twofold: First, in the early 1990s, the IDF's Advocate General upgraded the international law unit, turning it into a department headed by a full colonel. The department's greatest challenge has been to ensure that the military abide by the laws of war, to point to the legal aspects of the IDF's doctrine and operational planning, and to approve or prohibit the use of methods, such as targeted killing, or weapon systems under debate. The department has also been in charge of maintaining relations with international organizations, both governmental and non-governmental.

Second, since 1995, military lawyers, for the first time, became deeply involved in operational aspects. [84] During the 2006 Lebanon War, the 2009 Operation Cast Lead, the 2012 Operation Pillars of Defense, and the 2014 Operation Protective Edge, IDF legal advisers were already embedded in ground forces, ensuring real time compliance with international law. In January 2010, it was reported that Chief-of-Staff Gabi Ashkenazi had issued an order

---

83 Michael L. Gross, "Fighting by Other Means in the Mideast: A Critical Analysis of Israel's Assassination Policy," *Political Studies* Vol. 51, No, 2 (2003), pp. 350–68; Michael L. Gross, "Assassination: Killing in the Shadow of Self-Defense," in J. Irwin (ed.), *War and Virtual War: The Challenge to Communities* (Amsterdam: Rodopi, 2004), pp. 99–116; Mordechai Kremnitzer, *Are All Actions Acceptable in the Face of Terror? On Israel's Policy of Preventive (Targeted) Killing in Judea, Samaria and Gaza* (Jerusalem: Israel Democracy Institute, 2005) [Hebrew], pp. 11–13; David Kretzmer, "Targeted Killing of Suspected Terrorists," in *The European Journal of International Law*, Vol. 16, No. 2 (2005), p. 203; *USA Today*, 22 August 2006 <http://www.usatoday.com/news/world/2006–08–22-amnesty-lebanon_x.htm>; "Why They Died: Civilian Casualties in Lebanon during the 2006 War," *Human Rights Watch* 19 (2007) <http://www.reliefweb.int/rw/RWB.NSF/db900SID/SODA-76SB6T?OpenDocument>; "Fuelling Conflict: Foreign Arms Supplies to Israel/Gaza," <http://www.amnesty.org/en/library/asset/MDE15/012/2009/en/5be86fc2–994e-4eeb-a6e8–3ddf68c28b31/mde150122009en.html>; "Precisely Wrong," <http://www.hrw.org/en/news/2009/06/30/israel-misuse-drones-killed-civilians-gaza>.

84 Nathan Jeffay, "IDF Lawyers Now Coming Under Fire for Their Counsel During Gaza Conflict," *Forward*, 25 March 2009 <http://www.forward.com/articles/104254; <http://dover.idf.il/IDF/News_Channels/bamahana/07/43/01.htm>; <http://www.jpost.com/servlet/Satellite?cid=1233304648549&pagename=JPost%2FJPArticle%2FPrinter>.

requiring the IDF to consult with the army's legal advisers while military operations were underway and not merely at the planning stages. However, in order to prevent the legal advisers from disrupting the combat, the IDF decided that they will work only with the divisional headquarters rather than with brigade or battalion headquarters, as is common in some Western armies, including the US military.[85] It was also reported that greater emphasis had been placed on training officers in the rules of war and international law, as part of officer training courses at the level of company, battalion and brigade commanders, and that the IDF and Israeli Foreign Ministry had been tightening their cooperation in regards to their interactions with foreign governments and international organizations in order to ensure the legality of operations.[86]

The more legal advisers are involved in operational matters the greater the chances that operational considerations be subordinated to legal ones, to the point of degenerating operational sophistication and freedom of action. No wonder, therefore, that the Winograd Commission's Report on the management of the Second Lebanon War expressed concern over the growing reliance on legal advice in the course of military operations, which it considered liable to shift the responsibility from commanding officers to advisers and to divert commanders' attention from their operational challenges.[87]

Conclusion

Since the mid-1980s, Israel has experienced a type-of-war change, which reflected a shift from a reality of HICs to the ascendancy and pervasiveness of LICs. LICs have also become a "strategic threat," and have required adapting to the new reality at the levels of both military thought and practice. This change has been accompanied by some negative and positive aspects. One positive aspect has been a gradual understanding that the era of *Blitzkrieg* was over and that Israel will experience blow-for-blow confrontations that may last years and will deeply involve Israeli society. The Intifadas in particular triggered an intellectual effort to understand the nature of asymmetrical conflicts and attrition situations. Another positive development has been a more balanced approach to the role played by offense and defense, as compared to the ultimate commitment to offense in the past.

85 Anshel Pfeffer, "IDF to Seek Legal Advice during Future Conflicts," *Haaretz*, 6 January 2010.
86 Ibid.
87 Halutz's testimony before the Winograd Commission, Chapter 14, paragraphs 29 and 31 of the report.

Among the negative aspects one can point to the belief that decisive victory on the battlefield was no longer feasible (a belief that proved wrong in the First Lebanon War), and to the penetration of imported concepts from the US, such as the notion of "leverages and effects," and post-modern ideas, such as a "victory show;" and the underlining of LICs as wars of conviction, which underestimated the importance of and the need for real, material confrontation on the battlefield.

A second systemic factor has been technological developments that have shifted the dominance from maneuver to firepower, and have created a strong commitment to technology to the point of creating a cult of technology. The traditional, maneuver-oriented force multipliers have been replaced by firepower-oriented ones, e.g., offense and indirect approach carried out via fire and concentration of fire, instead of concentration of forces, etc. The idea of decisive victory via firepower and from the air has gained more and more supporters, culminating in the early stages of the Second Lebanon War. RMA-style doctrines have been adopted, and notions such as "information dominance," "dominant maneuver," "focused logistics," and "precision strike weapons" have been wrongly interpreted as enabling decisive victory with very low casualties and collateral damage, and the achievement of strategic results. Technology has also accounted for NCW, creating the illusion of the availability of perfect information of anything that happens on the battlefield. This, in turn, has increased the danger that once such information will not be available the troops will become almost blind. A new, technology-inspired conception of command and control has appeared, according to which commanders can run the battle not by leading their troops on the battlefield, but from their headquarters. Finally, the introduction of cyber warfare has pushed the IDF to employ it before having the chance to become fully acquainted with its theoretical and doctrinal aspects.

The third systemic factor has been legal and moral international constrains. These constrains have accounted for the sensitivity found in Israeli military thought and practice to the need to use force only as an act of self-defense; in ensuring that first strike is carried out only in such context; and that while using force the principles of discriminate use of force and proportionality must be kept, as well as the civil rights of enemy civilians. Since the mid-1990s, military lawyers have become involved in operational matters. Their mission has been to ensure real time compliance with international law, but it has increased the danger of degenerating operational sophistication and freedom of action. In reality, though, the military legal system often authorized operations and adopted relatively flexible interpretations of the law in order to justify IDF operations. Finally, greater emphasis has been placed on training officers in the rules of war and international law, as part of officer training courses at the level of company, battalion and brigade commanders.

CHAPTER 4

Unit (State)-Level Factors

This chapter analyzes unit-level factors, each accounting for a different aspect of Israeli military thought. First, Israel's geostrategic conditions, which have accounted for its "cult of the offensive;" second, cultural factors; third, social factors. Fourth, the penetration of self-imposed legal and moral imperatives into Israeli military thought stemming from Jewish and Israeli values to the point of its "judicialization," and the adoption of a post-heroic policy as a means of bridging the need for operational effectiveness, on the one hand, and morality on the other.

Geostrategic Factors and the Cult of the Offensive

Israel's tiny size; its location in the midst of an Arab war coalition on more than one front; and the imperative of achieving decisive victory as quickly as possible, for economic and societal reasons and in order to preempt superpower involvement and/or the arrival of Arab expeditionary forces upon the scene of war – have all led to the conclusion that Israel should avoid absorbing enemy attacks on its territory.

The practical conclusion of these conditions has been that Israel would have to transfer the war to the enemy's territory as quickly as possible[1] and to compensate for its quantitative inferiority by opting for offense as a major force multiplier. The offensive approach was first applied after the first truce in the War of Independence, took on a *Blitzkrieg* form in the 1956 Sinai War, and reached its peak in the 1967 Six-Day War. Beside its strategic benefits, offense was perceived as reflective of the IDF's spirit of initiation, audacity, improvisation and maneuverability, and as continuing the pre-State offensive nature of the *Haganah*. Given these and the fact that offense proved effective until 1967, it is no wonder that offense became second nature to the IDF. Among its staunchest advocates one can find political and military decision makers, both of the Right and Left wing, such as David Ben-Gurion, Menahem Begin, Yitzhak

1 Levite, *Offense and Defense in Israeli Military Doctrine*, p. 42; David Ben-Gurion, *Behilahem Israel* (Tel Aviv: Am Oved, 1975) [Hebrew], p. 90.

 | DOI 10.1163/9789004306868_005

Rabin, Yigal Allon, Ariel Sharon, Israel Tal, Dov Tamari, Shaul Mofaz, and others.

The negative aspect of the offensive approach was the unequivocal commitment to applying it, to the point of developing a cult of the offensive. The commitment had been so strong that when a basic change took place in Israel's strategic conditions as a result of the post-1967 borders, it failed to reevaluate the logic behind this approach. Although the new borders were considered "defensible borders," as compared to the pre-1967 "Auschwitz borders," to cite Foreign Minister Abba Eban,[2] and enabled Israel to opt for a defensive posture, Israel stuck to its offensive approach, falling victim to it in 1973.

Since the 1980s, a more balanced defensive/offensive approach has developed as a result of systemic changes in the type of war and technology rather than changes at the unit-level factors. These changes were addressed in Chapters 2 and 3.

Cultural Factors

A performance-oriented approach and experience-based intuition, as well as extolling resourcefulness and improvisation, have accounted for lack of interest in the intellectual aspect of the military profession. The post-1967 hubris has also been a major reason for the IDF's intellectual feebleness. Other cultural factors consisted of the difficulty to replace the tendency to think in terms of HIC norms and practices by LIC ones, and vice versa, and the negative impact of police missions in the territories.

"Bitzuism" (a performance-oriented approach) and experience-based intuition

"Sabra" is the common term for a native-born Israeli. It denotes the prickly pear cactus, which has a tough exterior but is soft and tender on the inside. The Sabras were considered a "new type of Jew," the antithesis of the "Diaspora Jew." One of their negative characteristics, however, has been the underestimation of the need to be well-educated, which is a prerequisite for advanced thinking. This was described by Abraham Shlonsky, one of Israel's greatest poets: "Yesterday, in the Diaspora, [... the Jews] were perfectly educated, much better than any educated gentile. Today, in the Land of Israel, they are flauntingly ignorant, simple and vulgar."[3]

---

2 *Der Spiegel*, 5 November 1969.

3 Oz Almog, *Sabra: The Creation of a New Jew* (Tel Aviv: Am Oved, 1997) [Hebrew], p. 218.

A research study conducted by Israeli sociologist Oz Almog on the Sabras pointed in part to their practice-oriented mentality. And indeed, once the Jews had a state and an army of their own, many Sabras felt that they were able to seize their fate in their own hands. When it came to problem solving, *bitzuism* – an approach that cherishes performance – became the dominant feature in many fields, both public and private, including the military.[4] Confident in their understanding of what should be done, and having succeeded in delivering the goods, the IDF's Sabra commanders often wondered why they had to study something that they were already adept at. This mindset, however, is not unique to the IDF. In reference to the American military thinking, Gregory Foster wrote: "It is ironic and disappointing that virtually all the reputed 'experts' on strategic and military affairs familiar to the public are civilian academicians, consultants and journalists."[5]

Since the 1960s and 1970s, the relatively few non-Sabra IDF commanders, some of whom were interested in the intellectual aspects of the military profession, gradually disappeared from the officer corps, leaving the stage almost completely to the performance-oriented Sabras. As the years passed, IDF commanders accumulated much experience, both in HICs and LICs, developing experience-based *coup d'oeil*.[6] Outstanding commanders during the Israeli War of Independence, like Yigal Allon and Yitzhak Rabin, were to a great extent a reflection of the Sabra mentality. As they could neither read nor speak English, they were hardly exposed to classical and modern texts. Their good performance should therefore be attributed to common sense, intuition, and experience rather than any intellectual preoccupation. Tuvia Ben-Moshe showed how the IDF's strategic and operational planning during the War of Independence had been the fruit of local experience and constraints and based on trial and error, rather than any theoretical understanding.[7] And Liddell Hart's biographer, Brian Bond, presented Yigal Allon's implementation of the indirect approach during Operation Horev on the southern front as an

4 Dan Horowitz and Moshe Lissak, *Trouble in Utopia: The Overburdened Policy of Israel* (Tel Aviv: Am Oved, 1990) [Hebrew], pp. 145, 181; Yaacov Hisdai, "Ideologue versus Performer: IDF's Priest and Prophet," *Maarachot* 279–280 (May-June 1981), pp. 41–6.

5 Gregory D. Foster, "Research, Writing, and the Mind of the Strategist," *Joint Force Quarterly*, 11 (Spring 1996), p. 115.

6 On the concept, see, for example, Frederick II (the Great), The King of Prussia's Military Instruction to his Generals, Article VI, "Of the Coup D'Oeil," <http://www.kw.igs.net/~tacit/artofwar/frederick.htm#VI>; Clausewitz, *On War*, p. 102.

7 Tuvia Ben-Moshe, "Liddell Hart and the Israel Defense Forces: A Reappraisal," *Journal of Contemporary History*, Vol. 16, No. 2 (April 1981), pp. 369–91.

expression of good intuition: "[Like Moliere's M. Jordain,] he had acted more or less instinctively, and then discovered that he had executed a brilliant indirect approach."[8] Bond also cited Rabin, who had admitted that indirect approach operations during the War of Independence had not been a result of pre-planned, Liddell Hart-based strategy or tactics. They just "largely coincided with Israel's choice of methods designed to overcome her inferiority in arms and numbers and the vulnerability of her people and territory."[9] Allon's biographer, Anita Shapira, portrayed him as the ultimate Sabra, who only years after having retired from the military service, was exposed to the intellectual aspects of strategy.[10] According to military historian Uri Milstein, the IDF's tactics have been based on the commanders' personal experience, whose education tracks have been based on mythos rather than any professional analysis and lesson-learning.[11] Brigadier-General (retired) Amatzia Chen argued that IDF officers have been so fully preoccupied with practice that they have had no awareness of the importance of military knowledge and thinking.[12]

Leutenant-Colonel Israel Beer criticized Israeli common belief that "even if elsewhere there exists a linkage between theory and practice, in our special case things are different. The IDF was born out of nothing, and its successes have been merely a result of the practical experience of practical soldiers." On the other hand, Beer gave pre-State and Israeli commanders the credit for having in the back of their mind the theory, doctrines and experiences of other militaries. He claimed that the operations of Orde Wingate's joint British-Jewish counterinsurgency unit in the second half of the 1930s in Palestine, which had a tremendous impact on Jewish pre-State military thinking, had been inspired by classical biblical doctrines and Oliver Cromwell's views, and that Yitzhak Sadeh's intellectual basis was etched in the Red Army's military thought. Beer argued also that the 1936–9 Arab Revolt had urged pre-State commanders to deepen their understanding of guerrilla warfare, which had caused them to turn to the British manual for India as well as the *Reichswher*'s education and training books, Lawrence of Arabia's writings, and so forth, and to study the linkage between World War II military events and the theoretical and doctrinal writings of Seeckt, Fuller, de Gaulle and Liddell Hart. Finally, Beer believed that the reprisal operations of the 1950s had followed in the

8 Brian Bond, *Liddell Hart: A Study of His Military Thought* (London: Cassell, 1977), p. 25.

9 Ibid., p. 246.

10 Anita Shapira, *Yigal Allon, Native Son: A Biography* (Philadelphia: University of Pennsylvania Press, 2007).

11 Uri Milstein, television interview (5 February 2012)<https://www.youtube.com/watch?v=hZZc4pKdLaE>.

12 Uri Milstein, "The IDF's March of Folly," *Nativ*, Vol. 116 (May-June 2007), pp. 46-54.

footsteps of the pre-State *Posh* (Field Companies) and *Palmach* operations, which in turn had been inspired by raids carried out by Gideon, David, and Judas Maccabeus, and that the battlefield successes of the IDF during the Sinai War had been inspired by World War II Western Desert battles. Despite all the above, it seems that Beer exaggerated the IDF's intellectual preoccupation, and that his analysis just reflected wishful thinking.[13]

The fact that good intuition often led IDF commanders to good decisions and success on the battlefield only strengthened the feeling that one could do without formal education. For example, unlike Moltke or Schlieffen, who despite their admiration for Clausewitz's teachings in general and his preference for defense in particular became offense and indirect approach advocates after having intellectually analyzed the changes that had taken place on the battlefield as a result of the Industrial Revolution, IDF commanders have been completely unaware of the whole spectrum of opinions presented by military thinkers on the question of which form of war, offense or defense, was stronger, that is, required less troops to fulfill the missions and entailed less casualties compared to the opposite form.[14] They, however, rightly chose offense as their preferred form of war due to Israel's aforementioned geostrategic conditions. Thanks to good, experience-based intuition and despite the cult of the offensive, when it came to LICs, the IDF was wise enough to apply a more balanced, offensive/defensive approach, as required by the nature of the challenge.[15] This point was elaborated in Chapter 3.

Extolling Resourcefulness and Improvisation

In contrast to their Arab opponents, whose image has traditionally been one of inflexibility on the battlefield that has limited their ability to exploit opportunities,[16] IDF commanders have demonstrated high adaptability to changing conditions on the battlefield, which has been related to as "improvising" skills. Moshe Dayan, chief-of-staff in the mid-1950s, pointed to the difference between the IDF and the Egyptian army which was exposed during

13 Israel Beer, "Theory and Practice in the Military Profession," *Maarachot* 130 (August 1960), pp. 26–7, 53.

14 Opinions have ranged from defense advocates (e.g., Clausewitz), a mixed approach proponents (e.g., Bernhardi, Liddell Hart) and offense advocates (e.g., Napoleon, Jomini, Moltke, Schliefen, Fuller).

15 Avi Kober, *Israel's Wars of Attrition: Attrition Challenges to Democratic States* (New York: Routledge, 2009).

16 Norwell De Atkine, "Why Arabs Lose Wars?" *Middle East Quarterly*, Vol. 6, No. 4 (December 1999), pp. 17–27; Kenneth M. Pollack, *Arabs at War: Military Effectiveness 1948–1991* (Lincoln: University of Nebraska Press, 2002).

the 1956 Sinai War: the Egyptians operated schematically, whereas the Israelis were more flexible and used to operating in a less conventional manner.[17] In a description of the fighting in the 1967 Six-Day War, Edward Luttwak and Dan Horowitz praised the Israeli commanders' "exceptional command flexibility," their ability "to change plans in the midst of combat," and "to improvise order out of the chaos of war," in contrast to the Arab commanders, who rigidly stuck to detailed, pre-prepared plans, and became confused once these plans were disrupted.[18] Tuvia Ben-Moshe likewise attributed good Israeli performance on the battlefield to pure improvisation.[19]

In the 1980s, Colonel (retired) Emanuel Wald identified a causal relationship between improvisation and lack of professionalism and anti-intellectualism in the IDF's general staff. "'[The general staff] is not academia' has been many general staff members' reaction to highly professional papers presented by a handful of skillful professional officers, who eventually discovered that their promotion was blocked. Praising improvisation became a self-defense mechanism, which compensated for lack of professionalism," Wald diagnosed.[20] More recently, the Winograd Commission pointed to the correlation between improvisation – which became part of the Israeli military culture – and lack of professionalism.[21] And a study by Dima Adamsky showed how Israel's anti-intellectual, practice-oriented strategic culture had accounted for its being the first to use RMA technology on the battlefield but very late to develop a conceptual framework for using it.[22]

Against this backdrop it is easier to explain the campaign held by the Chief of the IDF's Military Colleges, General Gershon HaCohen, against studying military history and theory. "Isn't it possible that [Chief-of Staff during the Sinai War, Moshe] Dayan was able to produce such a fascinating [operational] plan, precisely because he did *not* have to spend four years in studying Jomini and Clausewitz?" asked HaCohen rhetorically.[23] At the same time, he admitted

---

17 Moshe Dayan, *Story of My Life* (Tel Aviv: Idanim, 1976)[Hebrew], p. 244.

18 Luttwak and Horowitz, *The Israeli Army*, pp. 288–9.

19 Ben-Moshe, "Liddell Hart and the Israel Defense Forces."

20 Emanuel Wald, *Kilelat Hakelim Hashvurim*]The Curse of the Broken Vessels] (Tel Aviv: Shoken, 1987) [Hebrew], p. 183.

21 The Winograd Commission's final report <http://www.vaadatwino.org.il/pdf/לאינטרנט%20סופי%20מאוחד.pdf>, p. 425.

22 Dima Adamsky, *The Culture of Military Innovation* (Stanford: Stanford University Press, 2010).

23 Gershon HaCohen, "Educating Senior Officers," in *Is the IDF Prepared for Tomorrow's Challenges?* BESA Colloquia on Strategy and Diplomacy 24, July 2008, p. 30.

that General Norman Schwarzkopf, whose brilliant operational plan in the 1991 Gulf War he could not but praise, "had learned a lot" prior to that war,[24] something that Schwarzkopf himself was very proud of, happy to explain how a combination of the principles of war and Hannibal's indirect approach vis-à-vis the Romans during the Punic wars served as his main source of inspiration.[25]

Three examples from the 1973 October War can demonstrate the extent to which resourcefulness and improvisation became a source of flexibility under difficult combat conditions, although they were usually used to minimize the price for entering a bad situation rather than avoiding it in the first place. The first example is the quick recovery of Israeli ground troops from the surprise of the Egyptians' anti-tank warfare. Not only did Israeli tank crews apply new tactics of evading anti-tank missiles within a short period of time, they also dealt with the challenge in an inter-corps effort. Once they were joined by infantry and artillery, the problem was solved.[26]

The second example is from the air campaign of that war. The need to operate against the attacking Arab ground forces first, before achieving air superiority, countered IAF doctrine, according to which assisting the ground forces should take place only after the skies were cleared. The IAF, however, soon recovered from the difficulties it faced in handling the multi-layer Egyptian and Syrian surface-to-air missiles, thanks to a combination of quick lesson learning and doctrinal adaptations, suppression equipment that arrived from the US, and the use of ground forces in destroying SA missile batteries (about one third of the SA bases were destroyed by ground forces).[27]

The third example pertains to the difficulty of circumventing the concentration of Egyptian forces deployed along the Suez Canal due to the lack of sufficient crossing equipment. In the absence of this option, the IDF soon identified the opportunity to penetrate the Egyptian dispositions through the gap between the 2nd and 3rd armies, to cross the Canal, and reach relatively open spaces in the Egyptian forces' rear. The IDF only waited for the right timing to carry out this plan, which came once the Egyptian counterattack on October 14, 1973 failed and weakened the Egyptian ground forces significantly.

The tension between reliance on resourcefulness and improvisation, on the one hand, and using theoretical and doctrinal tools in order to cope with

24 Ibid., p. 32.

25 General Norman Schwarzkopf interviewed in "Hannibal and Desert Storm," *Timewatch* BBC Television, 1996.

26 Meir Finkel, *On Flexibility* (Tel Aviv: MOD, 2007) [Hebrew], pp. 194–207.

27 Ibid., pp. 208–22.

challenges on the battlefield, on the other, has been manifested in the IDF's command and control system. The IDF has been credited with a mission command system, which assigns the objectives or missions to be achieved, the zones of operation, and the time frame, but leaves the method to accomplish the missions to each commander's discretion. A good decentralized mission command is not supposed to be based on improvisation, but rather on a thorough educational and training process, during which all commanders acquire the same set of professional tools, which they will later apply according to the specific conditions on their particular battlefields. In the Prussian army, first under Moltke and then under his successors, superiors had confidence in their subordinates, knowing that having acquired common theoretical and doctrinal tools, each of them would act exactly as his superior would have acted had he been in his place.

The IDF's mission command system developed without reference to foreign models. According to Martin Van Creveld, even had it been inspired by the German *Auftragstaktik*, this could not be acknowledged overtly due to the sensitivities entailed in referring to any German system.[28] The IDF's mission command is attributed to former Chiefs-of-Staff Moshe Dayan in the mid-1950s and Yitzhak Rabin in the pre-1967 years.[29] Former Chief-of-Staff Mordechai Gur presented the two alternative command and control systems from which the IDF had to choose:

> A proper command system should be able to set itself goals and then strive to attain those goals in spite of the clear realization that things may go wrong, but when they do, the system is supposed to overcome [the problem]. There are two different options for such a system. The first is to plan everything in detail and only then start moving. The second is to define general objectives and start moving at once. The system will then gain momentum, and [missing] data will be supplied as progress is being made. The IDF chose the second of these ways. It is like a smart bomb being released on the basis of general data, without the target even being seen. Later on, after a few miles, it identifies the target and locks onto it. As of this point, it flies accurately until the target is acquired.[30]

28 Van Creveld, *The Sword and the Olive*, p. 169.

29 Ibid.

30 Mordechai Gur, "The IDF: Continuity versus Innovation," *Maarachot*, 261–262 (March 1978), pp. 4–6.

A mission command, however, proved to be inefficient at times, when commanders lacked resourcefulness. For example, in 1973, Chief-of-Staff David (Dado) Elazar lost faith in the resourcefulness and discretion of the O.C. Southern Command, appointing former Chief-of-Staff General (retired) Haim Bar-Lev as front commander.

A centralized control of the political and military echelons goes against the notion of mission command, violating the principle of exploiting success, as happened in 1982, when political limitations were one of the reasons for the IDF's failure to reach the Beirut-Damascus highway – a key operational center of gravity – within 48 hours.[31] During low-intensity operations in the territories, diplomatic and moral sensitivities have often imposed close control on the IDF by the political and military echelons, limiting its freedom of action. The RMA state of mind, too, has negatively affected the mission command system. As was previously argued, conducting battles from rear headquarters and over plasma screens, as happened during the Second Lebanon War,[32] instead of leading troops on the battlefield, was alien to Israeli mission command.

The Post-1967 Hubris and Intellectual Feebleness

Following the overwhelming exhibition of its military capabilities in 1967, Israel's siege mentality waned. With the signing of the peace agreement with Egypt, the sense of inferiority in the balance of forces was also tempered. Conversely, hubris on the Israeli side, as a result of the "aura of prestige" gained in the 1967 War, negatively affected Israeli decision makers and military.[33] According to Friedrich Nietzsche, "war makes the victor stupid."[34] And indeed, military achievements often result in complacency, as happened to Prussia after the Seven Years' War, to France in the wake of World War I, and to Israel in the wake of the 1967 Six-Day War. According to Van Creveld, "in retrospect, the smashing victory of 1967 was probably the worst thing that ever happened to Israel."[35] Referring specifically to Israeli military thought, General (retired) Avraham Rotem diagnosed: "it was rather a greater and stronger Israel that

31 Wald, *The Curse of the Broken Vessels*, p. 41.

32 Shelah and Limor, *Captives in Lebanon*, p. 385.

33 Eliot Cohen and John Gooch, *Military Misfortunes: The Anatomy of Failure in War* (New York: The Free Press, 1990), p. 124; Hanoch Bartov, *Dado* (Tel-Aviv: Maariv, 1978) [Hebrew], Vol. I, pp. 235–6; Professor Avi Ravitzky, Interview to *Maariv* for Israel's 53rd Anniversary <http://www.nrg.co.il/online/archive/ART/138/748.html>.

34 Friedrich Nietzsche, *Human, All Too Human* (Cambridge: Cambridge University Press, 1996), p. 163.

35 Van Creveld, *The Sword and the Olive*, pp. 198–9.

demonstrated [intellectual] feebleness. Both boldness and clarity of thought have disappeared."[36] And indeed, having just proven how good one had been on the battlefield, why should the Israeli bother to invest in thinking, learning, deliberating, innovating, or changing?

Slow Adaptation to a LIC Mindset

Although LICs have accompanied Israeli security environment since the early 1950s, until the mid-1980s the IDF tended to focus its military thinking on symmetrical HICs, a tendency that has not changed even against the backdrop of the decreasing pervasiveness and diminishing weight of HICs in the Arab-Israeli conflict. This had a few explanations, which seem to apply to militaries in general, and should be looked for not merely at the systemic level but also at the unit (state) level. First, militaries in general are often better at symmetrical, "regular" war, or at least feel most experienced with this type of warfare.[37] Second, states, and more so militaries, are used to considering symmetrical challenges to be more threatening than asymmetrical ones.[38] Israel's differentiation between "basic security" challenges, which stemmed from Arab regular armies, and "current security" ones that were considered a minor challenge, reflected such hierarchy of threats.[39] Third, symmetrical, "regular" wars are usually associated with greater buildup budgets.[40] This has been demonstrated by the fact that the lion's share of Israeli defense expenditures has been allocated to HIC challenges.[41] Fourth, HIC contexts usually lie at the heart of

36 Avraham Rotem, "Is a Small and Smart Military a Vision or a Legend?" in Golan (ed.), *Israel's Security Web*, p. 92.

37 Martin Van Creveld, "Less than Meets the Eye," *Journal of Strategic Studies*, Vol. 28, No. 3 (June 2005), p. 450; Robert M. Cassidy, *Russia in Afghanistan and Chechnya: Military Strategic Culture and the Paradoxes of Asymmetric Conflict* (Carlisle: US Army War College, Strategic Studies Institute, 2003) <http://www.smallwarsjournal.com/documents/russia.pdf>.

38 Ashton Carter, "Responding to the Threats: Preventive Defense," paper presented at the conference on 'Challenges to Global and Middle East Security,' Jaffee Center for Strategic Studies and Belfer Center for Science and International Affairs, Herzeliah, pp. 15–16 June 1998; *America's National Interests* (Washington, DC: Commission on America's National Interests, 1996). This report distinguishes between "vital," "extremely important," "just important," and "less important" national interests.

39 Shimon Peres, *The Next Phase* (Tel-Aviv: Am-Hassefer, 1965)[Hebrew], p. 11.

40 Deborah Avant, *Political Institutions and Military Change: Lessons from Peripheral Wars* (Ithaca: Cornell University Press, 1994), pp. 29–36, 117–29.

41 Kober, *Israel's Wars of Attrition*, Ch. 5.

armies' professional identity.[42] Fifth, at least until recent decades, "classical," regular war-oriented military thought was simpler. Sixth, the IDF, like military organizations in general, has suffered from entrenched traditionalism,[43] which has made it difficult to adapt to a new reality of war.

Had it not been for the 1987 First Intifada, the process of switching perception from one that focuses on HICs to another focusing on LICs would have been much slower. The following sections describe the gradual internalization that the role played by LICs in the Israeli-Palestinian conflict deserved a new thinking. As mentioned in Chapter 3, it was Defense Minister Yitzhak Rabin who, four months after the outbreak of the 1987 First Intifada, realized that Israel's military might was almost irrelevant to handling civil disobedience, and that LIC challenges could not be overcome in one attempt but rather via a cumulative process of exhaustion. In 1995, during his second tenure as Defense Minister, Rabin "upgraded" the LIC challenge, declaring for the first time that for Israel, terrorism represented a "strategic threat."[44]

During the First Intifada, the IDF's Central Command tested a new LIC doctrine in the territories,[45] the lessons of which were later applied during the Second Intifada of 2000. Prime Minister Ariel Sharon, who shortly after assuming power had changed former Prime Minister Ehud Barak's guideline to the IDF for handling the Intifada from "reducing" the violence to "ending" it,[46] soon adapted to a reality of attrition, which is typical to LICs. Unlike Chief-of-Staff Shaul Mofaz, who strove for decisive victory, both Sharon and Mofaz's successor Moshe Yaalon came to believe in winning the Second Intifada through a blow-for-blow battle stretching over years or maybe even decades.

When asked why, in contrast to his career as military commander he now refrained from striving for a speedy, decisive victory, Sharon explained that as prime minister one has to see the entire picture, not merely the narrow military one, and one must also assess how the situation might develop in the future, which requires a gradual approach.[47] "We have to control our actions in order to avoid escalation. This requires lots of perseverance, determination

42 Theo Farrell, "World Culture and Military Power," *Security Studies*, Vol. 14, No. 3 (September 2005), pp. 448–88.

43 Basil H. Liddell Hart, *Thoughts on War* (London: Faber & Faber, 1943), p. 30; Terry Terriff and Theo Farrell, "Military Change in the New Millennium," in Theo Farrell and Terry Terriff (eds), *The Sources of Military Change: Culture, Politics, Technology* (Boulder: Lynne Rienner, 2002), p. 265. Tony Mason, "Innovation and the Military Mind," <http://www.au.af.mil/au/awc/awcgate/au24-196>.

44 *Yediot Aharonot*, 30 January 1995.

45 Arieh O'Sullivan, "What a Riot," *Jerusalem Post*, 4 January 2004.

46 Amos Harel, "Zero Tolerance," *Haaretz*, 7 April 2006.

47 Carolyn B. Glick, "An Interview with PM Ariel Sharon," *Jerusalem Post*, 26 September 2002.

and peace of mind. One must understand that we cannot achieve everything now. It will take time [...], but at the end of the process the citizens of Israel will again feel secure. It will neither be short nor simple [...]."[48]

Yaalon for his part talked about a war that will be won by scoring points, as opposed to a knockout.[49] Already as Deputy Chief-of-Staff, he had described the confrontation with the Palestinians as one of an attritional nature: "[...The Palestinians] have identified the Israeli civilian rear's [low] staying power as a point of weakness, and have been focusing on it. [...] This confrontation is an important test of our staying power. If we cannot withstand it, it will impact on our relations with the Palestinians and the Israeli Arabs and on the way the Arab world sees us."[50] *Haaretz*'s military commentator, Zeev Schiff, described Chief-of-Staff Yaalon's view on attrition: "Yaalon believes that a guerrilla organization can be defeated in a prolonged war of attrition. It will not be a knockout, but a victory through points."[51]

The internalization of the fact that Israel's wars were now of a low-intensity, attritional nature became evident in declarations issued by Israeli officials during the Second Lebanon War. Prime Minister Ehud Olmert said it was "a day-to-day struggle, in which we ought to project cold bloodedness and determination."[52] He praised the Israeli citizens' staying power, depicting it as an important part of Israeli power.[53] Chief-of-Staff Dan Halutz said the rear's backing of the IDF was equal in importance to military power.[54] Deputy Chief-of-Staff Moshe Kaplinsky spoke of a war that will last weeks, although he did not predict it would take months to end.[55] Former IAF Chief General (retired) Eytan Ben-Eliyahu pointed to the societal staying power test as the decisive element in the confrontation, praising the Israeli rear's perseverance as something the government could lean on during the crisis.[56] IAF Chief Eliezer

48 Ben Kaspit, Interview with Ariel Sharon, *Maariv Weekend Supplement*, 13 April 2001.

49 Aluf Benn, "In Israel: Too Much to Leave to the Generals," *The Washington Post*, 18 August 2002.

50 Oded Granot and Itzik Saban, Interview with Moshe Yaalon, *Maariv Weekend Supplement*, 16 February 2001.

51 Zeev Schiff, "The Foresight Saga," *Haaretz*, 18 August 2006.

52 *Yediot Aharonot*, 16 July 2006.

53 Olmert's Speech before the Knesset, 17 July 2006 <http://www.pmo.gov.il/PMO/Archive/Speeches/2006/07/speechknesset170706.htm>.

54 <http://www.ynet.co.il/articles/0,7340,L-3278255,00.html>.

55 <http://www.nytimes.com/2006/07/18/world/middleeast/18cnd-mideast.html?ex=1310875200&en=922e5a114f378782&ei=5088&partner=rssnyt&emc=rss>.

56 Eitan Ben-Eliyahu, "The Rear Will be the Decisive Factor," *NRG News*, 16 July 2006 <http://www.nrg.co.il/online/1/ART1/449/981.html>.

Shkedy said the rear's staying power was the most effective weapon in Israel's hands during the confrontation.[57] And Minister of Transportation (former Defense Minister) Shaul Mofaz praised the Israeli rear's staying power, reminding the Israeli people of Sheikh Nasrallah's spider web theory and the expectation of the terrorist organizations that the morale of the Israeli civilian rear would break down within a short period of time.[58]

In mid-November 2006, following a Hamas or Islamic Jihad Qassam rocket attack from Gaza on the town of Sderot, Prime Minister Olmert delivered a speech on the difficulties entailed in coping with terror challenges. Terrorism cannot be wiped out with one blow, he said, reminding those who were demanding a "Defensive Shield"-like operation in Gaza that terror from the West Bank continued despite that operation.[59]

Lack of Institutional Intellectualism

Back in the 1950s, Yigal Allon offered a recipe for ensuring a satisfactory level of intellectualism among commanders: a mixture of formal education, which combines military and non-military studies (the latter in order to widen the commander's horizons), and supplementary education completed independently, according to each officer's own will and intellectual thirst.[60] Indeed, ideally, the drive to develop and maintain intellectual curiosity, to appreciate history-based military theory, and to believe in its practical dividends should come from within the military. The Prussian/German army, since the early 19th century,[61] the Soviet military, both during the interwar period and the Cold War,[62] and some of the leading American, British, and Australian military colleges[63] – have all shared the belief that the commander's intellectual knowledge should and can be transformed into capability, and that military history and

57 <http://www.ynet.co.il/articles/0,7340,L-3276824,00.html>.

58 <http://www.nrg.co.il/online/1/ART1/449/775.html>.

59 Barak Ravid, "Olmert: In This War There is No Once and For All.," <http://www.nrg.co.il/online/1/ART1/506/249.html>.

60 Yigal Allon, *Curtain of Sand* (Tel Aviv: Hakibbutz Hameuhad, 1968) [Hebrew], p. 303.

61 Herbert Rosinski, *The German Army* (New York: Praeger, 1966).

62 Adamsky, "The Conceptual Co-Influence."

63 Williamson Murray, "The Army's Advanced Strategic Art Program," *Parameters*, Vol. 30, No. 4 (Winter 2000–01), pp. 31–39; Michael Evans, "From the Long Peace to the Long War: Armed Conflict and Military Education and Training in the 21st Century," <http://www.defence.gov.au/jetwc/docs/publications%202010/PublcnsOccasional_310310_FromtheLongPeace.pdf>.

theory are the foundations of doctrine, planning and practice. This is reflected in the curricula of middle and senior rank training programs of institutions such as the School of Advanced Military Studies (SAMS); the Marine Corps' School of Advanced Warfighting (SAW); or the National Defense University (NDU), particularly the National War College (NWC), and the College of International Security Affairs (CISA). If a military fails to acknowledge the need to develop and encourage "intellectual soldiers" in an institutionalized manner, it is the duty and responsibility of the political echelon and Parliament (in the Israeli case, the Knesset) to impose systematic military thinking that lives up to the standards of the best militaries in the world, as had been done in the Soviet Union by Mikhail Gorbachev in the mid-1980s,[64] or by the Canadian government in the mid-1990s.[65]

When the IDF's Command and Staff College and Tactical Command College were established in 1954 and 1999 respectively, it seemed to mark a breakthrough in the institutionalized professional education and training of Israeli commanders. Unfortunately, the IDF has remained skeptic with regard to the contribution of the intellectual aspects of the military profession, particularly military theory. Its reluctance to invest in these aspects was blatantly expressed in many instances throughout the years. Chapter 2 mentioned a few examples of such an attitude: the suspension of the publication of classical theory books by *Maarachot*; the absence of mandatory reading lists in the IDF; the closure of the *Barak* program at the Command and Staff College; the suspension of the publication of some IDF professional journals such as *Maarachot*; or the Chief of the Military Colleges' objection to the study of military history and theory.

Politization and Militarization Processes

Horowitz and Lissak have described two opposing processes that have been typical of the relations between the military and society in Israel: a politization of the military and a militarization of the Israeli society and leadership.[66] And indeed, politization existed particularly during the state's early years. There were military units with political affiliation, like the *Palmach*; officers were allowed to openly be members of political parties, particularly if they belonged

64 Jennifer G. Mathers, "Reform and the Russian Military," in Farrell and Terriff (eds), *The Sources of Military Change: Culture, Politics, Technology*, pp. 161–84.

65 David J. Bercuson, "Up from the Ashes: The Re-Professionalization of the Canadian Forces After the Somalia Affair," in Stuart Cohen (ed.), *The New Citizen Armies* (London: Routledge, 2010), pp. 159–69.

66 Dan Horowitz, and Moshe Lissak. *Trouble in Utopia: The Overburdened Policy of Israel* (Tel Aviv: Am Oved, 1990) [Hebrew].

to the "correct" ruling elite; and nominations of senior commanders were affected by political considerations. Prime Minister David Ben-Gurion tried to put an end to this phenomenon, in the hope of making the IDF more professional and free of political and ideological influence. His efforts have been crowned with success. But even during the days of politicization Israeli military thought was hardly affected by political considerations.

A more problematic phenomenon has been the militarization of politics, a process that has become part of Israeli strategic culture. For many years now the military echelon has had enormous influence over decision-making processes and policymaking on issues of war and peace. The political echelon has traditionally preferred the IDF as its staff for security matters, and the military intelligence has been in charge of early warning at the national level. The special status of the IDF can be explained by its image as highly professional, experienced and impartial, and highly obedient, even in situations of disagreement with the political echelon on matters such as going to war, war objectives, war plans, etc. This explains, at least partially, why the IDF has enjoyed a monopoly over Israeli military thought and why the civilian authorities have refrained from interfering in purely military thought or from imposing greater intellectualization on the military.

Societal Factors

An Occupational – Rather than Institutional – Military

During Israel's early decades, the *Kibutzim*, the collective communities where many of the older military elite's commanders hailed from, were against any long-term professional education for officers, not only because for socialists military professionalism had a fascist, militaristic connotation, but also because that might have induced officers to choose a military career and eventually leave the *Kibbutz* for good. This, too, played a negative role in commanders' professionalism and explains why many of them based their functioning on their personal experience instead.[67]

The process wherein the Israeli army gradually became an occupational – rather than institutional – organization[68] opened up an opportunity for greater investment in the professional education of IDF commanders. However, a growing number of military personnel whose professions can be found outside

67 Uri Milstein, television interview (5 February 2012)<https://www.youtube.com/watch?v=hZZc4pKdLaE>.

68 Charles C, Moskos, "Institutional and Occupational Trends in Armed Forces: An Update," *Armed Forces & Society*, Vol. 12, No. 3 (1986), pp. 377–82.

the military, many of whom have just happened to work for the military, seems to have pushed the IDF further away from military intellectualism and professionalism.

New Elites and Their Impact of the IDF's Military Thought

As was pointed out in Chapter 2, during Israel's formative years (1948–56) there was an exceptional interest in the intellectual aspects of the military profession, and the 1950s and the 1960s saw the development of extensive doctrinal thinking. This is partially explained by the natural thirst for professional inspiration during these challenging years. Commanders such as Yigael Yadin, Yohanan Ratner, and Yehoshua Globerman, had become acquainted with European military thought out of intellectual curiosity. These individuals could be considered "military intellectuals." Another partial explanation for this phenomenon is foreign influence. Although the IDF has never consciously selected its commanders by their social group affiliation,[69] during the state's early years a hard core of the IDF commanders and staff officers were foreign-language speakers of European origin, who were exposed to British, German, Austrian, or Russian military thought. Some of them had served in European armies prior to World War II or during that war, e.g., Haim Laskov in the British army, and Eytan Avisar (formerly named Sigmund von Friedman), Rephael Lev, and Israel Beer in the Austrian army.

The impact of the latter group was compatible with Prime Minister Ben-Gurion's effort to build the IDF as a "Western Army." Conversely, the skills of young Israelis originating from Arab states were considered disparagingly similar to those typical of soldiers of Arab armies. A document issued by the IDF's Manpower Branch in the early 1950s depicted them as being "slow and easy going, absorbing very little from the world and contributing very little to it." The document argued that their "low intelligence and past habits made it very difficult for them to live up to their new society's requirements." They were also described as suffering from "a strong sense of inferiority." The officers' selection processes gave preference to candidates born in Israel or in the US and Europe. In the period 1959–61, for example, the chances of a Sabra or a US and Europe born candidates to be sent to officers' training school were 8.5 percent and 5.4 percent respectively, as compared to only 1.4 percent for candidates originating from Arab countries.[70]

It was only after the 1973 October war that this picture started changing. The old social elite's share in the officer corps was gradually diminishing. This

69 Reuven Gal, *A Portrait of the Israeli Soldier* (Westport: Greenwood, 1986), p. 116.

70 Amir Oren, "Afeka, not Ofakim," *Haaretz*, 17 September 2010.

happened for a number of reasons, such as the doubling of the IDF's size; the eroding image of the military after 1973, 1982, and the Intifadas; more opportunities opened for young educated Israelis in a free market economy (see below); and the closing of the educational gap between the old elite and the formerly peripheral groups.

Whereas in the past, up to 30 percent of the commanders came from the ideologically-motivated *Kibbutzim* and *Moshavim* – although these collective or cooperative settlements comprised no more than eight percent of the Israeli population in the early years, and three percent and even less in later periods[71] – in the more recent decades an increasing number of commanders have come from the geographical and sociological periphery. According to Yagil Levy, the new "ethno-national coalition" was led by religious Zionists and Sepharadi Jews, who were later joined by immigrants from the former Soviet Union. The former group, whose members' representation in the IDF's Infantry rose from 2.5 percent in 1990 to 31.4 percent in 2007 – filled the ideological void that had been created by the disappearance of the old elite, though with a different content. For the second and third groups, the military has served a social mobilization tool.[72] Levy argues that given the erosion of the IDF's image and status, there was no real competition between the old and new elites over dominance in the military. The new elites were climbing a "descending escalator" that no longer interested the old elite. An "occupational orientation" has not yet fully taken over, and there are still discerned islands of an "institutional orientation" among them, which used to characterize previous generations of IDF commanders. Commanders from the new elite are not less intelligent than their predecessors, and many of them have received a good education. But it does not appear that they have brought with them any significant change, for better or worse, in the inclination or capacity for intellectualism.

The Near-Lost Competition with the Civilian Sector

Another social explanation for the poor intellectualism of the military seems to be its competition with the civilian sector. As long as Israel retained mandatory military service, the IDF benefited from the availability of middle and

71 Luttwak and Horowitz, *The Israeli Army*, p. 184; Gal, *A Portrait of the Israeli Soldier*, pp. 81–3.

72 On the new coalition, see Yagil Levy, "Materialist Militarism," *Alternative Information Center*, 8 August 2006 <http://www.alternativenews.org/news/english/materialist-militarism-20060808.html>; Yagil Levy, *Israel's Materialist Militarism* (Lanham: Lexington Books, 2007); Yagil Levy, "The War of the Peripheries: A Social Mapping of IDF Casualties in the Al-Aqsa Intifada," *Social Identities*, Vol. 12, No. 3 (May 2006), pp. 309–24.

higher-middle class soldiers. This, however, was balanced by the tendency of many high-quality conscripts to leave the service after three years in order to start a civilian career with higher materialistic prospects and without the risk of constant mortal danger.

In the contemporary highly-materialistic Israeli society, the most sought-after careers are as businesspeople and high-tech developers, rather than as farmers, warriors, or intellectuals. The fact that a military career in Israel comes to its end once servicemen have reached their forties has imposed on most of them a second career. This has dissuaded well-educated young Israelis from the old and new elites alike, who were more creative and intellectually-equipped for analytical and critical thinking, from opting for a military career in the first place, prompting them to prefer a civil career from the start.

An inner survey conducted by the IDF in the late 1960s – at its heyday– pointed to a correlation between officers' quality and their tendency to prefer a civilian career.[73] In November 1980, former IAF Chief General Benny Peled complained that too many talented servicemen had decided to leave the military for civilian jobs.[74] Another survey, which was conducted in the 2000s, reaffirmed this trend. More than 50 percent of the IDF's junior officers and NCOs considered leaving service and starting a civilian career.[75] According to the IDF's Manpower Department, in 2011 12.2 percent of the IDF's officers and NCOs serving in top technological units, such as the Computer Service Directorate, the Israeli NSA unit 8200, or cyber warfare units, left the army for a civilian career, where they could double or triple their income. The number grew to 17.3 percent in 2012 and to 21 percent in 2013. A similar trend was depicted among qualitative field commanders who were marked as candidates for military career: in 2011 29.3 percent of them chose to leave the military, and the number grew to 32.5 percent in 2012 and to 37.9 percent in 2013.[76]

In an attempt to keep skillful commanders in the military, the IDF has enabled many of them to receive academic degrees in fields that would help them in their second career. Former Chief-of-Staff Ehud Barak even made an ideology of it, explaining that as a rule, an academic education, in whatever field, would make better commanders. This approach, however, ignored the danger that this might widen the gap between the acquired academic education and the military profession's needs. Given this policy, which was

73 <http://www.kav.org.il/100994/921>.

74 Interview to the monthly *Nihul* – Israel's Management Magazine, in November 1980.

75 *Maariv*, 12 October 2007.

76 Yossi Yehoshua, "Brain Damage to the IDF," *Yediot Aharonot Weekend Supplement*, 12 February 2014.

implemented by Barak's successors, a recommendation by a Knesset Committee headed by MK Brigadier-General (retired) Ephraim Eitam to establish a military academy seemed to be swimming against the current.

Explanations for Israeli Post-Heroic Mindset's First Rule

As pointed out in Chapter 2, post-heroic warfare has become a dominant feature of Israeli way of war, particularly in asymmetrical LICs. Post-heroic warfare has two major rules. The first rule dictates that one is not "allowed" to get killed, whereas the second rule is that one is also not allowed to kill innocent enemy civilians. Most explanations for post-heroic warfare focus on its first rule. In his writings from the 1990s, Edward Luttwak preferred a realist, demographical explanation that put the finger on the poor birth rates in Western democracies.[77] But it seems that the most compelling explanation for post-heroic warfare at the unit level is a greater casualty aversion in conflicts that do not involve the vital interests, let alone the survival, of Western democracies. The lesser the vitality of interests, the lesser the readiness to sacrifice for their achievements and the greater the preference by governments and citizens for low-cost, low-risk operations.[78] Technological developments, particularly the ability to target enemies from afar, to use unmanned platforms, or to deploy active defense systems, as well as precision weapons, and a very short sensor-to-shooter cycle that has made it possible to kill a target shortly after it has been identified and to avoid the killing of noncombatants who otherwise might have been on target-site during the attack – have all played a facilitating role in implementing post-heroic warfare.

Commitment to Jewish and Western Democratic Moral and Legal Standards

Moral and legal considerations have had a considerable impact on Israeli military thought, not merely in the form of a systemic constraint but rather as a self-imposed commitment to Israeli democratic principles, law, and institutions; Jewish tradition; and universal moral values.[79] All these have permeated Israeli military thought to the point of becoming an integral part of it. The public debate held in Israel in the wake of collateral damage allegedly inflicted

---

77 Luttwak, "Post-Heroic War."

78 Ibid.

79 "The IDF's Spirit" (Code of Ethics) <http://dover.idf.il/IDF/About/Purpose/Code_Of_Ethics.htm>.

by IDF troops, the deeply-rooted norm of the purity of arms, the formulation of a code of ethics by the IDF, the occasional rules by the Israeli Supreme Court, and the absence of any incidents of sexual violence against Arab women by IDF troops – have all been indicative of the Israeli commitment to fight as morally as the conditions have allowed, despite the fact that for its enemies the end has usually justified the means. A tension has developed between law and ethics, on the one hand, and the need to ensure operational effectiveness, on the other, and effort has been made to bridge the two, by adopting a post-heroic way of war.

"Purity of Arms"

The guiding principle in Israeli military thought and practice that has served as a moral lighthouse has been called the "purity of arms." The purity of arms norm has existed since pre-State times, constituting a moral obligation to refrain from hurting innocent civilians during military operations. More specifically, it meant that Israeli troops were forbidden to kill noncombatants, loot or rape – even with the display of such conduct by the enemy – because it has been considered immoral.

The basic notion of purity of arms was worded by Zionist socialist ideologue Berl Katznelson during the Arab Revolt in Palestine (1936–9). Katznelson said it was imperative that arms were not used against noncombatants, "so that our weapons will not be stained with the blood of the innocent."[80] David Ben-Gurion, already a leading figure at the time, also expressed his objection to the killing of innocent civilians, but unlike Katznelson, his reasoning was not solely based on moral considerations, but also on utilitarian reasons, namely, the expected criticism against the Jewish community.[81] Yet, the policy of restraint, including the purity of arms, which was part of it, was rejected by the National Military Organization (*Etzel*), a militant Jewish underground.

The purity of arms principle was later adopted by the *Palmach* – the *Haganah*'s strike force – as one of its ethical principles, and was sometimes attributed to Yitzhak Sadeh, the *Palmach*'s commander.[82] It is noteworthy that in 2002, IAF Chief Dan Halutz considered the purity of arms a fundamentally

80 Wikipedia, "The Restraint Policy," <http://he.wikipedia.org/wiki/%D7%9E%D7%93%D7%99%D7%A0%D7%99%D7%95%D7%AA_%D7%94%D7%94%D7%91%D7%9C%D7%92%D7%94>.

81 Menachem Finkelstein, "The Purity of Arms," <http://www.daat.ac.il/mishpat-ivri/skirot/235-2.htm>.

82 For the views among the Jewish community's leadership during the Arab Revolt regarding the policy of restraint and the moral issues related to it, see Yaacov Shavit, *Self-Restraint or Reaction* (Ramat Gan: Bar-Ilan University Press, 1983) [Hebrew].

invalid concept, arguing that weapons were never intended to be pure and there is no such thing as immaculate wars. According to Halutz, there is only an appropriate and proper use of force, which preserves one's human image.[83]

Justifying the Use of Force

Just War

The IDF's name implies that its destiny has been to defend the State of Israel. Any use of force intended to serve this purpose would be justified in Israeli eyes as an act of self-defense. This has also been compatible with the Jewish notion of obligatory war (*Milhemet Mitzvah*, in Hebrew).

Reflective of the importance of abiding by the principle of just war was the debate held in Israel between revisionists and status-quo protagonists. The status-quo school was closer to the notion of just war, which was for years the dominant approach to war in Israel. According to the status-quo school, domestic legitimacy to war could be ensured only if the direct defense of Israel was at stake. Once attacked, however, it would be legitimate to seize the opportunity and to achieve more ambitious objectives,[84] as was the case before and during the 1967 War, when Israel's narrow, vulnerable borders were considered "Auschwitz borders." A representative of this school was Israel Tal.[85]

Israeli revisionists, too, did not ignore nor reject the need to fight just wars. Unlike the competing school, however, they argued that a threatened nation had the right to initiate war in order to create a more convenient territorial and/or political strategic environment.[86] Prime Minister Menahem Begin, who held such a view, was not alone in this opinion. He shared this view with Israeli leaders such as right-wing politician Rehav'am Zeevi. "Soft" revisionists rejected the idea of initiating war, but, on the other hand, justified the occasional adoption of ambitious political war objectives, in case that war was forced upon Israel. For example, Yigal Allon, a retired General and Deputy Prime Minister, thought that in 1967 Israel missed the opportunity to reach Cairo, Damascus, and Amman in order to put an end to the bloody Arab-Israeli conflict.[87] Yuval Ne'eman, a senior reserve Intelligence officer and later on one of Israel's leading scientists, argued that in the case of an Arab attack, Israel

---

83 Vered Levy-Barzilai, "The High and the Mighty," *Haaretz*, 21 August 2002.

84 Yitzhak Rabin, *Pinkas Sherut* (Tel-Aviv: Maariv, 1979) [Hebrew], p. 22.

85 Israel Tal, *National Security: The Few against the Many* (Tel-Aviv: Dvir, 1996) [Hebrew], pp. 54–56.

86 Yehezkel Dror, *A Grand Strategy for Israel* (Jerusalem: Academon, 1989) [Hebrew], pp. 165–73.

87 Yigal Allon, *Kellim Shluvim* (Tel-Aviv: Hakibbutz Hameuhad, 1980) [Hebrew], p. 111.

should feel free to consider the possibility of shattering the Syrian state, establishing an independent Druze state in the Horan (a geographic area located in southwestern Syria), encouraging the establishment of a Druze state in Northern Iraq, and annexing Southern Lebanon and the strategically-important Edom Mountains.[88] It seems that the heightened public debate in Israel with regard to the legitimacy of the revisionist "war by choice" during the 1982 First Lebanon War and in its aftermath only strengthened the status-quo school of thought.

It is also noteworthy that Israel has refrained from initiating LICs because they have been perceived as a type of war that has played into the hands of its Arab enemies, preferring *Blitzkrieg* instead. As LICs have been imposed on it, they have been considered compatible with the notion of just war.

The Legality and Morality of Targeted Killing

During the Second Intifada, killing military operatives and political leaders was extensively used by Israel as a counter-terror strategy, raising questions pertaining to both self-defense and the right of prevention and/or preemption. Israeli and foreign scholars, as well as Israeli Attorney General and the Military Advocate General, defended Israel's right to carry out targeted killing, considering it an act of "active self-defense," provided that all other measures had failed, and based on the premise that the killing was aimed at those who had terrorized the lives of innocent civilians. As such, targeted killing could be considered a form of preemption.[89] Even the supporters of this counter-terror method on moral grounds have made clear that it should not be used as a

---

88 Efraim Inbar, *The Outlook on War of the Political Elite in the Eighties* (Jerusalem: The Leonard Davis Institute for International Relations, 1988), p. 23.

89 Amos N. Guiora, "Targeted Killing as Active Self-Defense," *Case Western Reserve Journal of International Law* 36 (2004), pp. 319–34;" Asa Kasher, "The Morality of Preemptive Warfare," *Maariv*, 12 January 2001; Steven R. David, *Fatal Choices: Israel's Policy of Targeted Killing* (Ramat-Gan: BESA Center for Strategic Studies, 2002); Emanuel Gross, "Self-defense against Terrorism: What Does It Mean? The Israeli Perspective," *Journal of Military Ethics*, Vol. 1, No. 2 (2002), pp. 91–108; Emanuel Gross, *The Struggle of Democracy Against Terrorism: Lessons from the United States, the United Kingdom, and Israel* (Charlottesville: University of Virginia Press, 2006), pp. 220–46; Daniel Statman, "Targeted Killing," *Theoretical Inquiries in Law*, Vol. 5, No. 1 (January 2004), pp. 179–98;" Ward Thomas, "The New Age of Assassination," *SAIS Review*, Vol. 25, No. 1 (Winter– Spring 2005), p. 34; Abraham D. Sofaer, "Response to Terrorism: Targeted Killing is a Necessary Option," *San Francisco Chronicle*, 26 March 2004 <http://www.sfgate.com/cgi-bin/article.cgi?file=/chronicle/archive/2004/03/26/EDGK65QPC41.DTL>; Gal Luft, "The Logic of Israel's Targeted Killing," *The Middle East Quarterly* 10 (2003) <http://www.meforum.org/article/515>; Gideon Alon, "Mofaz: IDF Jurist Approves Killings," *Haaretz*, 11 January 2001.

means of vengeance or punishment, or for achieving side benefits such as deterrence, or to be applied on a much larger scale than originally intended, beyond a "ticking bomb" scenario,[90] restrictions that have not been fully observed in practice given the need to be as efficient as possible in a situation of war against terror.

Code of Ethics

In the absence of explicit guidelines regarding military operations taking place among civilians, particularly during the Intifadas, the IDF had to fill the lacuna. To this effect, it formulated an ethical code that was supposed to serve as a guiding light for its troops in general and in the territories in particular. The code was first published in 1995, and was then revised in 2000. The current incarnation of the purity of arms principle is a section in the Code of Ethics. It states that

> IDF servicemen and women will use their weapons and force only for the purpose of their mission, and only to the necessary extent, and will maintain their humanity even during combat. IDF soldiers will not use their weapons and force to harm human beings who are not combatants or prisoners of war, and will do all in their power to avoid causing harm to their lives, bodies, dignity and property.[91]

In early 2004, the IDF added a code of ethics for combating terrorism,[92] and distributed it among its field units and commanders. The code outlined rules of behavior during operational activity in the territories, calling on the troops to observe the rules "in order to maintain the image of a humane and ethical army."[93]

The "Judicialization" of Israeli Military Thinking

As a result of the Intifadas, greater emphasis has been placed on training commanders in the rules of war. At the same time, awareness has grown in Israel of the need to integrate moral and legal considerations into operational planning. The IDF's international law department (ILD) was instructed to ensure that the military abided by the laws of war, to point to the legal aspects of the IDF's

---

90 Amos Yadlin and Asa Kasher, "The Ethics of Fighting Terror," *Journal of National Defense Studies* 2–3 (2003), pp. 5–12.

91 The IDF's Code of Ethics.

92 *Yediot Aharonot*, 9 February 2004.

93 *Haaretz*, 10 March 2004.

doctrine and operational planning, and to either approve or prohibit the use of methods such as targeted killing, or weapon systems under debate.

As pointed out in Cahpter 2, between 1995 and 2004 military lawyers became involved in operational aspects, and Chief-of-Staff Gabi Ashkenazi issued an order requiring the IDF to consult with the army's legal advisers while military operations were underway and not just when they were being planned.[94] The more legal advisers were involved in operational matters the greater the chances that operational considerations would be subordinated to legal ones, to the point of degenerating operational sophistication, freedom of action, and operational effectiveness.

Conclusion

Israeli military thought has been affected by the following unit-level factors. First, the state's geostrategic conditions have accounted for the preference of offense to the point of developing a cult of the offensive. Second, cultural factors: A performance-oriented approach and experience-based intuition, extolling resourcefulness and improvisation, have accounted for the lack of interest in the intellectual aspect of the military profession; and the post-1967 hubris has been one of the major reasons for the IDF's intellectual feebleness. Other cultural factors have been thc difficulty to replace the tendency to think in terms of HIC norms and practices by LIC ones, and vice versa, i.e., the negative impact of police missions in the territories on the IDF's HIC thinking, and the lack of institutional intellectualism. Third, societal factors: the occupational orientation of the IDF; the impact of a new military elite; and the near-lost competition with the civilian sector over attracting well-educated and skillful young Israelis to a military career. Fourth, the growing role played by moral and legal constrains, stemming from Jewish and Israeli values.

Not only has the IDF developed a strong commitment to a moral conduct of war, which has nevertheless imposed serious constraints on military action, likened to fighting with one hand tied behind the back, but the moral and legal aspects have become an integral part of its military thought and practice. The growing reliance on legal advice before and in the course of military operations has created the danger that the commanding officers' attention would be diverted from their missions to the legal advice pertaining to these missions, a phenomenon that has gained the name Judicialization.

---

94 Anshel Pfeffer, "IDF to Seek Legal Advice during Future Conflicts," *Haaretz*, 6 January 2010.

CHAPTER 5

Conclusion

The Conclusion summarizes the factors that affect the intellectual and modern focus of military thought, and then offers a short account of the state of Israeli military thought, the factors that have shaped it, and some directions for improving it.

General Formative Factors

Military thought is affected by both realist and non-realist factors. For example, a realist explanation for *the cult of technology* would focus on the impact of technological developments. A non-realist explanation, on the other hand, would point to the culture of a particular nation or military, particularly to the tendency to value technical solutions to strategic problems. A combined explanation would treat these two explanations either as an independent variable or as an intervening variable, depending on the points of departure.

An interesting question is whether there is such a thing as universal culture. The answer to this question is that in most cases military culture is unique and differs from one case to another. At the same time, however, common traits can be found among groups of players, for example, the pre-World War I cult of the offensive that was shared by the central European great powers, or military organizations' tendency towards entrenched traditionalism.

Two systemic factors affect military thought: The nature of war, and *Zeitgeist*. One of the major changes in modern war and strategy pertains to technology. Technology has cast its long shadow on every aspect of modern war and strategy. Manifestations of its dominant role and its impact on military thought can be easily identified: the squaring of Clausewitz's triangle so that it would include technology, as the major representative of the non-material aspects of war; technology as an important dimension in Michael Howard's so-called "forgotten dimensions of strategy;" technology as the major reason for the emergence of the operational level-of-war; technology as a major explanation for the deepening of the battlefield and its expansion to the opponents' civilian rear; technology as the creator of the aerial dimension and later on the outer space and the cyberspace; and technological developments as accounting for the aforementioned *cult of technology*, particularly in Western countries.

 | DOI 10.1163/9789004306868_006

In recent decades, nonstate players have acquired technologies that in the past would have been found solely in the hands of regular armies. The availability of advanced technologies for the weaker adversaries, and the difficulty Western democracies often face when coping with terror and guerrilla challenges have created a new reality, which has narrowed the capabilities gap between state and nonstate players, blurring the borderlines between HICs and LICs, and serving as a major factor in the emergence of concepts such as hybrid war, or 4GW.

Finally, technology has been involved in all kinds of revolutions in war in modern time, both "small revolutions," as a result of the marriage of operational and technological depth processes, and "big revolutions" that are created by the marriage of technological and societal depth processes.

Another development that has had a major impact on military thought is the broadening of war. This process has manifested itself in various forms. One of them has been the receding of war from the direct battlefield and the involvement of entire societies in war. This brought Liddell Hart to formulate the concept of grand-strategy, and Michael Howard to point to the importance of the aforementioned forgotten dimensions of war. The broadening process has also created the need to add yet another level of war, the operational one. Unfortunately, the emphasis on the non-operational aspects of war and strategy has constituted a major reason for the neglect of the core aspect of strategy, i.e., the military one, as one can learn from the gradual shift of focus from "military studies" before World War II to "strategic studies" after World War II, and to "security studies" in recent decades.

The broadening of war has been accompanied by a parallel process of narrowing that has taken place for reasons such as the continued lowering of casualty rates during military confrontations, often as a result of an attempt – and a technological capability – to avoid the killing of one's own troops and of innocent civilians, which is typical of post-heroic warfare, or the reducing size of armies as result of the dramatic improvement of weapons' performance, their rising cost, and the need to minimize attrition rates on the battlefield. The notion of small but smart militaries has taken hold among highly-developed armies, as reflected by their adherence to RMA ideas, post-heroic warfare, etc.

The broadening and narrowing of war have not been alone in depicting the nature of war. The war of today can also be characterized by greater complexity. Examples include Hoffman and Mattis's notion of "Hybrid Wars," which reflects the obliteration of the borderline between HICs and LICs; Steven Metz's "Gray Area War," which refers to a situation that involves an enemy that seeks primarily profit, but which has political overtones and a substantially-greater capability for strategic planning and the conduct of armed conflict than traditional criminal groups; the blurred borderlines between military and police operations, one of its typical expressions being the phenomenon of narco-terrorism; or the emergence of what Chris Bellamy named the empty/saturated battlefield, which is emptier of forces, as a result of the aforementioned reasons, but is saturated with fire.

Another systemic factor has been the pervasiveness of LICs in the post-World War II period (80 percent of the conflicts in the international system). This process was hardly matched by greater interest in LICs, as military thinkers faced difficulties in reorienting their minds to a LICs reality. Once LICs eventually received the attention they deserved – which happened no earlier than the post-Cold War era, when they constituted some 95 percent of the conflicts – military thinkers repeated their former mistake, but this time the other way round: A wave of theoretical and empirical works on ethnic conflicts, such as Non-Trinitarian War (Martin Van Creveld) and State-to-Nation Balance (Benjamin Miller), to name a few, has swept the security studies discipline during recent decades. This wave was accompanied by a growing tendency towards post-modern notions, which was reflected in works on Post-Heroic Warfare (Edward Luttwak), Hybrid War (Hoffman and Mattis), 4GW (Thomas Hammes and William Lind), and others, or the emergence of notions such as "victory image."

As far as *Zeitgeist* is concerned, three aspects of it have affected military thought most, while weakening its external validity. First, military thinkers have had difficulty in abstracting themselves from the nature of war in a given period. Their works have often reflected war as seen through the eyes of people living in their own time, imparting to their military thinking a contemporary color. This phenomenon has manifested itself in a number of ways. For example, Clausewitz assigned paramount importance to events relating to his own time, and the Napoleonic wars were predominant in the formulation of his theory.

Second, military thought often reflects a temporary, specific intellectual climate. For example, prominent sources of inspiration that left their imprint on modern military thought were the ideas of the European Enlightenment during the 17th and 18th centuries and Romanticism in the 19th century. The Enlightenment provided military thought with a scientific and rationalistic patina, and manifested itself in the desire to formulate principles of war that would encompass military wisdom within the confines of a few general rules, at the risk of reducing the principles of war to a "manual" for the commander, similar to the laws in the natural sciences. Thinkers belonging to the principles-of-war school have been Frederick the Great, Henry the Duke of Rohan, the Marquise de Silva, Henry Lloyd, Jomini, Basil H. Liddell Hart, and others.

Romanticists like Clausewitz, on the other hand, rejected this prescriptive and doctrinaire approach to war, claiming instead that war refuses to be subordinated to a set of rules, and that it has a spiritual, emotional and intuitive side that not only requires the genius of the commander as a precondition for facing the uncertainties of war, but also a high morale and martial zeal among the troops.

Military thought between the Enlightenment and Romanticism reflected a blend of the two periods. For example, Jomini stood far closer to the Enlightenment than Clausewitz. Though admitting to the artistic nature of warfare, he believed that within the framework of this art it was possible to formulate a set of rules that contained a prescription for the successful conduct of the war (a notion adopted by Liddell Hart in the 20th century).

Third, as early as the interwar period, and much more so during the post-Cold War era, moral and legal aspects of war have become unprecedentedly important. The notions of just war, discriminate use of force, proportionality, and civil liberties have permeated military thought, particularly in Western democracies engaged in non-existential LICs.

At the unit (state) level, the association of military thought with players' specific strategic conditions rather than universal conditions has caused military thought to focus on doctrine rather than theory. Sometimes a doctrine for a specific player has gained the status of theory, as happened to Douhet's thought on airpower. Other formative factors at the unit level can be organizational, such as the tension between promotion aspirations and intellectualism; the tendency of the "military mind" towards conservatism and entrenched traditionalism; or the impact of lesson learning from past wars.

Individuals have affected military thought through three channels. First, many military thinkers have drawn their inspiration from famous military commanders of the past or earlier thinkers. For example, Jomini's ideas were inspired by those of the Welshman Henry Lloyd; Moltke and Schlieffen, as well as Marxist military thinkers, such as Engels or Mao, drew much inspiration from Clausewitz; and Mahan adopted many ideas regarding naval warfare from Jomini's theories on land warfare. Clausewitz stands out for the profusion of non-military thinkers who inspired him, including Plato, Aristotle, Montesquieu, Rousseau, Kant, and perhaps Hegel. He had a wide range of interests, which included art, literature, history and philosophy. These no doubt broadened his horizons and, besides sharpening his intellectual and analytical powers, deepened his insight into everything connected with social phenomena.

Second, the personal experience of thinkers. For example, those who had witnessed the weaknesses of insurgency, e.g., Engels in Germany or Moltke in

France, concluded that an equipped, disciplined and determined regular force could defeat insurgents. On the other hand, those who had witnessed insurgents' successes, such as Jomini (based on his experience in Spain), or Lawrence of Arabia (based on his experience in the Arabian Peninsula), were much more optimistic regarding the chances of insurgents gaining the upper hand.

Third, thinkers' determination to overcome entrenched traditionalism on the part of the military organization of their country or among their fellow military men. Thanks to their personal curiosity, commitment to innovation, and courage to fight for their ideas many of them have become important thinkers despite the negative attitude towards their ideas, e.g., Tuchachevsky, Douhet, Mitchell and others.

The Israeli Case

The State of Israeli Military Thought: Intellectual and Modern Focus

Throughout the years, the Israeli military establishment has shown symptoms of poor intellectualism that have been reflected, among other expressions, in a lack of interest in the theoretical aspects of the military profession and the underestimation of theory's contribution to practice. This lack of intellectualism has had a detrimental effect on the IDF's performance, particularly during the First and Second Lebanon Wars and the First Intifada. Since the 1990s, the IDF has been emulating RMA-inspired American doctrine, at the expense of its originality and innovation. As if to add insult to injury, Israeli military thinking has been affected by false intellectualism and intellectual pretense.

These negative trends have been mitigated, though, by a number of positive symptoms, such as vibrant military thought during Israel's formative years, both before the establishment of the State of Israel and during its early years; great debates over force structure, organization, force design and doctrine throughout the years; a relative popularity of military history; or the establishment of the Command and Staff College in 1954 and the Tactical Command College in the 1990s.

As far as the modern focus of Israeli military thought is concerned, it has been characterized by four major developments. First, a late adaptation to LIC challenges, and an eventual exaggerated bent toward LIC challenges, which has raised questions regarding the IDF's preparedness for HICs. Second, a strong tactical orientation. A late adaptation to the operational and grand-strategic levels, however, has taken place since the 1980s/early 1990s, and a two directional relationship has emerged between the tactical and the strategic/grand strategic levels that has manifested itself in the "tacticization of grand-strategy" and the "grand-strategicization of tactics." Third, a strong technological orientation in recent decades, which has gradually become second nature, and has eroded the IDF's operational art. The repercussions of technology have included the erosion of Israel's traditional force multipliers as a result of the ascendancy of firepower over maneuver, and a different logistical logic as a result of the declining role played by maneuver. Finally, moral and legal considerations have become an integral part of Israeli military thought.

A comparison of *Maarachot* and the American journal *Military Review* articles shows that as far as the modern focus is concerned the two journals reflect similar changes in war and strategy. When it comes to the intellectual focus, however, the impression one gets from the comparison is that Israeli military thought lags behind American military thought.

Formative Factors

The Systemic Level

Since the mid-1980s, Israel has experienced a change in the type of war from a HIC reality to a LIC reality. LICs have become a major Israeli "strategic threat," and have required adapting to the new reality both in terms of thought and practice. And indeed, the IDF has gradually internalized that the era of *Blitzkrieg* was over and that Israel will experience blow-for-blow confrontations that may last years, instead. The Intifadas, in particular, have triggered an intellectual effort to understand the nature of asymmetrical conflicts. The LIC reality has also imposed a more balanced approach to offense and defense, as compared to the ultimate commitment to offense in Israeli HICs. Other negative aspects of the new reality have been the belief that decisive victory against nonstate players would be impossible to achieve, and the penetration of imported concepts from the US, such as the notion of "leverages and effects," and post-modern ideas, such as "victory image" as a substitute for the material/physical aspect of military confrontation.

A second systemic factor that has cast its shadow on Israeli military thought has been technological developments. Generally speaking, the IDF has shifted from a balanced view of the role played by technology to a cult of technology. The ascendancy of firepower over maneuver, particularly the role played by PGMs, has diverted the attention from maneuver to firepower. An RMA-style doctrine has been adopted, which was founded on the baseless belief that success on the battlefield could be achieved via the use of firepower and/or from the air, instead of ground maneuver. "Information dominance," "dominant maneuver," "focused logistics," and "precision strike weapons" have all been believed to achieve rapid, decisive victory, particularly in HIC scenarios, with very low casualties and collateral damage, and strategic results. Network

Centric Warfare (NCW) has created the illusion that perfect, real-time information about anything that happens on the battlefield would be available, which is dangerous given the possibility that in the absence of such information, troops will have to operate almost blindly. Another negative development has been the beginning of a tendency to run battles not by leading the troops on the battlefield, but rather from headquarters. This has constituted a challenge to the IDF's traditional command and control system, which commanders would deny. Finally, the introduction of cyber warfare has found the IDF in a position of employing a new technology before having the chance to get acquainted with its theoretical and doctrinal aspects. Such warfare has already challenged traditional conceptions such as offense/defense, deterrence and early warning, as had happened years ago once nuclear weapons were introduced.

A third systemic formative factor has been legal and moral international constrains, which have accounted for the Israeli acknowledgement of the need to use force only as an act of self-defense; to make sure that preemption, let alone prevention, is carried out only in such a context; to abide by the principles of discriminate use of force and proportionality; and to honor enemy civilians' civil rights.

The Unit (State) Level

Four factors at the unit level have had an effect on Israeli military thought. First, the state's geostrategic conditions have accounted for the preference of offense to the point of developing a cult of the offensive. Second, cultural factors. "*Bitzuism*" and experience-based intuition, extolling resourcefulness and improvisation, have accounted for the lack of interest in the intellectual aspect of the military profession; and the post-1967 hubris has been one of the major reasons for the IDF's intellectual feebleness. Other cultural factors have been the difficulty to replace the tendency to think in terms of HIC norms and practices by LIC ones, and vice versa, i.e., the negative impact of police missions in the territories on the IDF's HIC thinking; and the lack of institutional intellectualism, which is to a large extent a result of the dominance of the IDF in security matters. Third, societal factors: the occupational orientation of the IDF; the impact of a new military elite; and the almost-lost competition with the civilian sector over the well-educated and skillful young Israelis who tend to shy away from a military career. Fourth, the growing role played by moral and legal constrains, stemming from Jewish and Israeli values.

The Level of the Individual

There have hardly been issues that Israeli individual military experts have not referred to. Their interest and activity have ranged from military history, structure and organization, force design, doctrine, planning, and so forth. Individuals have also led great debates on these issues. But decades of rich battle experiences have failed to produce either a significant number of Israeli generals of great worldwide reputation, or a significant number of renowned military theorists. The lack of acquaintance with classical military thought has often caused individuals to reinvent the wheel, as happened with the concept of attrition.

What can be done to improve the levels of intellectualism of Israeli military thinking?

War has been changing, and Israel's enemies have become more sophisticated. The IDF's experience has diminished, and too many military misfortunes in recent decades have exposed its insufficient professionalism. Being practical soldiers can no longer deliver the goods in terms of effectiveness. Against this backdrop, the ability to apply knowledge for solving practical problems would have an added value. It would also create professional pride among IDF commanders, strengthen their self-confidence, and strengthen their image in the eyes of society.

Ideally, the drive to develop and maintain intellectual curiosity, to appreciate history-based military theory, and to believe in its practical dividends should come from within the military. The Prussian/German army since the early 19th century, the Soviet military, both during the interwar period and the Cold War, or some of the leading American, British, or Australian military colleges – have all shared the belief that the commander's intellectual knowledge should and can be transformed into capability, and that military history and theory are the foundations of doctrine, planning and practice.

In order to produce Israeli "intellectual soldiers" a reform is required that would replace the IDF's intuition and experience-based practice with a knowledge-oriented approach. Military thought feeds best on an environment and atmosphere that encourage the preoccupation with the intellectual aspects of the military profession, and not on a handful of individuals, who are courageous enough and intellectually competent to develop ideas of great value in the military field against the background of entrenched traditionalism. A reform in the IDF can take place and succeed only if it undergoes a process of institutional intellectualism. In the case of a military which fails to acknowledge the need to develop and encourage "intellectual soldiers" independently, it is the duty and responsibility of the political echelon and Parliament (in the Israeli case, the Knesset) to impose systematic military thinking that lives up to the standards of the best militaries in the world, as had been done in the Soviet

Union by Mikhail Gorbachev in the mid-1980s, or by the Canadian government in the mid-1990s.

Once the necessity of a reform is acknowledged, it would have to be translated into practical terms. It is necessary that the IDF treat all stages of military education as parts of a structured whole. In the current educational system the lower-rank commanders programs emphasize the tactical and operational levels before any broad understanding of the nature of war is provided. It must be replaced with a top-down process, in the spirit of Scharnhorst's and Clausewitz's approach, that regarded theory as providing the concept of the "whole," that is, the systemic order which explains how each part of war relates to another. An education and training authority would be needed in order to enforce a significant labor division between the programs at all levels. At the same time, the IDF should strengthen its joint professional military education, as the basis of service interoperability, while maintaining the commitment to the principle of the unity of command.

The IDF had better dissolve its unholy alliance with the Israeli academia. It deserves its own military and security studies academic education, based on recruiting the best experts across the Israeli academia, instead of using every now and then an occasional university as the IDF's academic studies sponsor and courses provider. Mandatory reading lists for commanders would expose them to relevant abstract and practical professional literature. Instructors in military courses must live up to the highest intellectual standards.

The IDF could benefit most from following the example of three American institutions: the School for Advanced Airpower and Space Studies (SAASS), which educates talented middle-rank airpower officers who are considered potential strategists; the Army's School of Advanced Military Studies (SAMS); and the Marine Corps' School of Advanced Warfighting (SAW) – the latter two focusing on educating commanders at the operational level. Studies in SAASS are based on the art of war and historical inquiry. Each student is required to research and write a substantial original thesis, and the school instructors are encouraged to prove excellence not only in teaching but also in researching, writing and publishing scholarly articles and books. Although SAASS belongs to the Air Force, it views itself as a joint professional military institution. Its philosophy closely matches the one adopted by SAMS and SAW. One can also learn from British institutions such as the Joint Services Command and Staff College (JSCSC), which trains commanders and staff officers of all three UK Armed Services and officers from countries around the world. The JSCSC has a unique academic-military partnership with King's College London.

It would be most recommended that the IDF either adopt the philosophy of these institutions or send outstanding officers who have an excellent command of the English language to attend one year or even a six-month military program in the US or the UK. This would expose them to higher professional education standards, and get them up to speed in their professional knowledge.

It would be appropriate to conclude the book with the assertion that not only can a deep acquaintance with the abstract aspects of the military profession be a decisive factor in success on the battlefield, proficiency, not only in the practical aspects of the military profession but also in its abstract aspects, helps create professional pride among commanders and improves their image in the eyes of the society, and can be translated into strengthened self-confidence of commanders in their profession.

Appendix

As explained in the introduction, in order to find out if there is a correlation between the trends identified in *Maarachot* publications and those in journals abroad, or if these trends are unique to Israel, a partial comparison is made with the American journal *Military Review*. This journal shares similar orientation with *Maarachot*: Like *Maarachot* it is committed first and foremost to covering land warfare issues, but alongside this focus it is also interested in the wider picture, and in a variety of topics related to changes in the nature of war and its conduct. The comparison is limited to a selected period – the Cold War's late years and the post-Cold War period – during which dramatic changes in war and strategy have taken place.

The comparison shows that as far as the modern focus is concerned the two journals reflect similar changes in war and strategy. When it comes to the intellectual focus, however, the impression one gets from the comparison is that Israeli military thought lags behind American military thought. Below are some specific references to MR, with comparison to *Maarachot*.

(a) The representation of regulars among authors is much higher in *MR* than in *Maarachot*

TABLE A1 *Status 1983–1987*

	N	%
Regulars	393	70.18
Reservists	23	4.11
Retired	37	6.61
Civilians	71	12.68
Non-American	36	6.43
Total	560	100

TABLE A2 *Status 1988–1994*

	N	%
Regulars	538	65.37
Reservists	38	4.62

 | DOI 10.1163/9789004306868_007

TABLE A2 *Cont.*

	N	%
Retired	76	9.23
Civilians	143	17.37
Non-American/Anonym	28	3.4
Total	823	100

TABLE A3 *Status 1995–2000*

	N	%
Regulars	314	57.83
Reservists	30	5.52
Retired	80	14.73
Civilians	100	18.42
Non-American/Anonym	19	3.5
Total	543	100

TABLE A4 *Status 2001–2004*

	N	%
Regulars	208	54.74
Reservists	17	4.47
Retired	79	20.79
Civilians	61	16.05
Non-American/Anonym	15	3.95
Total	380	100

TABLE A5 *Status 2005–2008*

	N	%
Regulars	173	45.29
Reservists	12	3.14
Retired	66	17.28
Civilians	93	24.35

Non-American/Anonym	38	9.95
Total	382	100

(b) **In *MR*, authors holding the rank of Major constitute the most frequent group of authors, whereas in *Maarachot* Majors are only third, after Lieutenant-Colonels and Colonels.**[1]

TABLE A6 *Rank 1983–1987*

	N	%
General	7	1.25
Lieutenant-General	1	0.18
Major-General	8	1.43
Brigadier-General	4	0.72
Colonel	52	9.3
Lieutenant-Colonel	99	17.71
Major	158	28.26
Captain	59	10.55
First and Second Lieutenant	1	0.18
NCO	3	0.54
Non-regulars	167	29.87
Total	559	100

1 The ranks in the US are not always equivalent to those in the IDF. The rank of General exists in the Army (General 4 Stars), the Air Force and the Marines, and is equivalent to the rank of Admiral in the Navy. / The rank of Lieutenant-General exists in the Army (General 3 Stars), the Air Force and the Marines, and is equivalent to Vice-Admiral in the Navy. / The rank of Major-General refers to the Army (General 2 Stars), the Air Force and the Marines, and is equivalent to the rank of Rear-Admiral in the Navy. / The rank of Brigadier-General exists in the Army (1 Star), the Air Force and the Marines, and is equivalent to Commodore in the Navy. / The rank of Colonel exists in the Army, the Air Force and the Marines, and is equivalent to Captain in the Navy. / The rank of Lieutenant-Colonel refers to the Army and the Air Force and is equivalent to the rank of Commander in the Navy. / The rank of Major refers to the Army and the Air Force. Its equivalent in the Navy is Lieutenant-Commander. / The rank of Captain exists in the Army and the Air Force. It is equivalent to Lieutenant in the Navy. / The IDF has the ranks of Lieutenant-General, Major-General, Brigadier-General, Colonel, Lieutenant-Colonel, Major, Captain, Lieutenant, and Second Lieutenant.

TABLE A7 *Rank 1988–1994*

	N	%
General	27	3.28
Lieutenant-General	17	2.07
Major-General	21	2.55
Brigadier-General	15	1.82
Colonel	94	11.42
Lieutenant-Colonel	149	18.1
Major	162	19.68
Captain	46	5.59
First and Second Lieutenant	-	-
NCO	7	0.85
Non-regulars/Anonymous	285	34.63
Total	823	100

TABLE A8 *Rank 1995–2000*

	N	%
General	16	2.95
Lieutenant-General	6	1.1
Major-General	14	2.58
Brigadier-General	6	1.1
Colonel	70	12.89
Lieutenant-Colonel	85	15.65
Major	99	18.23
Captain	14	2.58
First and Second Lieutenant		
NCO	4	0.74
Non-regulars/Anonymous	229	42.17
Total	543	100

TABLE A9 *Rank 2001–2004*

	N	%
General	2	0.53
Lieutenant-General	6	1.58
Major-General	5	1.32
Brigadier-General	4	1.05
Colonel	35	9.21
Lieutenant-Colonel	60	15.79
Major	80	21.05
Captain	13	3.42
First and Second Lieutenant		
NCO	3	0.79
Non-regulars/Anonymous	172	45.26
Total	380	100

TABLE A10 *Rank 2005–2008*

	N	%
General	6	1.57
Lieutenant-General	10	2.62
Major-General	4	1.05
Brigadier-General	8	2.09
Colonel	25	6.54
Lieutenant-Colonel	39	10.21
Major	65	17.02
Captain	15	3.93
First and Second Lieutenant	-	-
NCO	1	0.26
Non-regulars/Anonymous	209	54.71
Total	382	100

(c) *MR* authors show more interest in military theory than *Maarachot* authors.

TABLE A11 *Military theory 1983–1987*

	N	%
Theory	100	20.28
Doctrine	175	35.50
Planning	60	12.17
No reference to the above topics	158	32.05
Total	493	100

TABLE A12 *Military theory 1988–1994*

	N	%
Theory	103	15.08
Doctrine	258	37.77
Planning	101	14.79
No reference to the above topics	221	32.36
Total	683	100

TABLE A13 *Military theory 1995–2000*

	N	%
Theory	48	10.88
Doctrine	171	38.77
Planning	55	12.47
No reference to the above topics	167	37.87
Total	441	100

TABLE A14 *Military theory 2001–2004*

	N	%
Theory	46	15.92
Doctrine	110	38.06
Planning	44	15.22
No reference to the above topics	89	30.8
Total	289	100

TABLE A15 *Military theory 2005–2008*

	N	%
Theory	46	15.65
Doctrine	84	28.57
Planning	59	20.07
No reference to the above topics	105	35.71
Total	294	100

(d) Authors of both journals share a considerable interest in military history.

TABLE A16 *Military history 1983–1987*

	N	%
History	209	42.39
No reference	284	57.61
Total	493	100

TABLE A17 *Military history 1988–1994*

	N	%
History	368	53.88
No reference	315	46.12
Total	683	100

TABLE A18 *Military history 1995–2000*

	N	%
History	232	52.61
No reference	209	47.39
Total	441	100

TABLE A19 *Military history 2001–2004*

	N	%
History	154	53.29
No reference	135	46.71
Total	289	100

TABLE A20 *Military history 2005–2008*

	N	%
History	186	63.26
No reference	108	36.74
Total	294	100

(e) As the time passed authors of both journals became much more preoccupied with LICs. In *MR* the preoccupation with LICs grew from 16 percent in the period 1983–1988 to 65 percent in the period 2005–2008. *Maarachot* publications, too, reflect growing interest in LICs over the years (from 1 percent in the period 1983–1987 to 29 Percent in the period 2005–2008), but also a much slower adaptation to a LIC reality.

TABLE A21 *Types of conflict 1983–1987*

	N	%
Unconventional	60	12.17
Conventional	308	62.47

Sub-conventional	80	16.23
No reference to conflict	45	9.13
Total	493	100

TABLE A22 *Types of conflict 1988–1994*

	N	%
Unconventional	39	5.71
Conventional	387	56.66
Sub-conventional	208	30.45
No reference to conflict	49	7.17
Total	683	100

TABLE A23 *Types of conflict 1995–2000*

	N	%
Unconventional	21	4.76
Conventional	176	39.91
Sub-conventional	202	45.8
No reference to conflict	42	9.52
Total	441	100

TABLE A24 *Types of conflict 2001–2004*

	N	%
Unconventional	8	2.77
Conventional	62	21.45
Sub-conventional	198	68.51
No reference to conflict	21	7.27
Total	289	100

TABLE A25 *Types of conflict 2005–2008*

	N	%
Unconventional	10	3.4
Conventional	24	8.16
Sub-conventional	255	86.73
No reference to conflict	5	1.7
Total	294	100

(f) Whereas *MR* authors were much more interested in the operational dimension of strategy, in *Maarachot* the technological dimension became the leading one.

TABLE A26 *Dimensions of strategy 1983–1987*

	N	%
Operational	191	38.74
Societal	32	6.49
Technological	61	12.37
Logistical	32	30.77
No reference to dimensions of strategy	177	35.9
Total	493	100

TABLE A27 *Dimensions of strategy 1988–1994*

	N	%
Operational	301	44.1
Societal	81	11.9
Technological	109	15.9
Logistical	61	8.9
No reference to dimensions of strategy	131	19.2
Total	683	100

TABLE A28 *Dimensions of strategy 1995–2000*

	N	%
Operational	171	38.77
Societal	59	13.38
Technological	120	27.21
Logistical	32	7.26
No reference to dimensions of strategy	59	13.38
Total	441	100

TABLE A29 *Dimensions of strategy 2005–2008*

	N	%
Operational	137	46.6
Societal	41	13.95
Technological	39	13.26
Logistical	20	6.8
No reference to dimensions of strategy	57	19.39
Total	294	100

(g) The interest in the operational level on the part of *MR* authors notwithstanding, it was only second to the interest in the strategic level, but is still quite high. In *Maarachot* the operational level was almost ignored in the 1983–1987 period but with the years it drew increasing attention.

TABLE A30 *Levels of war 1983–1987*

	N	%
Grand Strategy	7	1.42
Strategy	132	26.77
Operational level	94	19.1
Tactics	118	23.94
No reference to levels of war	142	28.80
Total	493	100

TABLE A31 *Levels of war 1988–1994*

	N	%
Grand Strategy	14	2.05
Strategy	233	34.11
Operational level	197	28.84
Tactics	186	27.23
No reference to levels of war	53	7.76
Total	683	100

TABLE A32 *Levels of war 1995–2000*

	N	%
Grand Strategy	7	1.59
Strategy	134	30.38
Operational level	157	35.6
Tactics	110	24.94
No reference to levels of war	33	7.48
Total	441	100

TABLE A33 *Levels of war 2001–2004*

	N	%
Grand Strategy	3	1.04
Strategy	106	36.68
Operational level	88	30.45
Tactics	84	29.06
No reference to levels of war	8	2.77
Total	289	100

TABLE A34 *Levels of war 2005–2008*

	N	%
Grand Strategy	3	1.02

Strategy	137	46.6
Operational level	92	31.29
Tactics	49	16.66
No reference to levels of war	13	4.42
Total	294	100

These findings reflect:

a. A greater respect for professionalism in general and the intellectual basis of the military profession in particular on the part of *MR* authors.
b. Earlier professional maturity of American officers holding the rank of Major, most probably as a result of a greater investment in military education.
c. Differences in threat perception in the US and Israel, as a result of which the IDF lagged behind the US in addressing LIC challenges, focusing for too many years on HICs.
d. A greater interest in the operational level among *MR* authors as a result of the development of AirLand Battle doctrine. The fact that the interest in the operational level on the part of *MR* authors is only second to the interest in the strategic level may be interpreted as an attempt to refer to war and strategy in a more holistic manner, something that was already recommended by Clausewitz.
e. A more balanced approach to the role played by technology in war and strategy on the part of *MR* authors, as compared to the leading role attributed to technology by *Maarachot* authors.

Bibliography

"The IDF's Spirit" (Code of Ethics) <http://dover.idf.il/IDF/About/Purpose/Code_Of_Ethics.htm>.

"Why They Died: Civilian Casualties in Lebanon during the 2006 War," *Human Rights Watch* 19 (2007).

Abulafia, Amir. "The Courage to Express Independent Opinions," *Maarachot* 433 (October 2010), pp. 20–7.

Adamsky, Dima P. "The Conceptual Co-Influence: the Soviet Military-Technical Revolution and the Western Military Innovations," paper presented at the IAIS annual meeting, Hebrew University, 6 June 2006.

———. *The Culture of Military Innovation: The Impact of Cultural Factors on the Revolution in Military Affairs in Russia, the US, and Israel* (Stanford: Stanford University Press, 2010).

Ahad Ha'am. *Essays, Letters, Memoirs* (Oxford: East and West Library, 1946).

Alberts, David S., John J. Garstka, and Frederick P. Stein. "Network-Centric Warfare: Developing and Leveraging Information Superiority," <http://www.dtic.mil/cgi-bin/GetTRDoc?Location=U2&doc=GetTRDoc.pdf&AD=ADA406255>.

Alger, John I. *The Quest for Victory: The History of the Principles of War* (Westport: Greenwood Press, 1982).

Allon, Amnon. "Teaching Generals," *Haaretz Weekend Supplement*, 2 August 2002.

Allon, Yigal. *A Curtain of Sand* (Tel Aviv: Hakibbutz Hameuhad, 1968) [Hebrew].

———. *Communicating Vessels* (Tel Aviv: Hakibbutz Hameuhad, 1980) [Hebrew].

Almog, Oz. *Sabra: The Creation of a New Jew* (Tel Aviv: Am Oved, 1997) [Hebrew].

Alon, Gideon. "Mofaz: IDF Jurist Approves Killings," *Haaretz*, 11 January 2001.

America's National Interests (Washington, DC: Commission on America's National Interests, 1996).

Aron, Raymond. *Peace and War* (New York: Anchor Books, 1973).

Arquilla, John and David Ronfeldt. *Swarming & the Future of Conflict* (Santa Monica: Rand, 2000).

———. *Swarming and the Future of Conflict* (Santa Monica: Rand, 2000).

Asprey, Robert B. *War in the Shadows: The Guerrilla in History* (New York: Morrow, 1994).

Assa, Haim and Yedidya Yaari. *Diffused Warfare* (Tel Aviv: Yediot Aharonot, 2005) [Hebrew].

Atkine, Norwell De. "Why Arabs Lose Wars?" *Middle East Quarterly*, Vol. 6, No. 4 (December 1999), pp. 17–27.

Attrition Strategy in a Limited Confrontation, A Symposium at the Israeli National Defense College, 10 January 2002.

Augustine, Norman Ralph. *Augustine's Laws* (New York: Viking, 1983), Law No. 16, 1984.

Avant, Deborah D. *Political Institutions and Military Change: Lessons from Peripheral Wars* (Ithaca: Cornell University Press, 1994).

Bamo, David W. "Military Intellectualism," House Armed Services Subcommittee on Investigations and Oversight, 10 September 2009.

Bar, Micha. "The Tank's Obscure Future," *Maarachot* 339 (February 1995), pp. 2–9.

Bar-Joseph, Uri. *The Watchman Fell Asleep: The Surprise of Yom Kippur and its Sources* (Tel-Aviv: Zmora-Bitan, 2001) [Hebrew].

Bar-Kochva, Moshe. *Chariots of Steel* (Tel-Aviv: Maarachot, 1989) [Hebrew].

Barnea, Nahum and Shimon Shiffer. Interview with Prime Minister Ehud Olmert, *Yediot Aharonot New Year Supplement*, 29 September 2008.

Bar-On, Mordechai. "The Sinai Campaign: Reasons and Achievements," *Skira Hodshit*, 10–11 (1986), pp. 4–16.

Bartov, Hanoch. *Dado* (Tel-Aviv: Maariv, 1978) [Hebrew].

Barzilai, Amnon. "[Chief of the IDF's Technology and Logistics Branch General Udi] Adam's Technological Revolution," *Haaretz*, 2 April 2004.

Batschelet, Allen W. *Effects-Based Operations: A New Operational Model?* (Carlisle: US Army War College, 2002).

Beer, Israel. "Conservatism and Flexibility in Military Thought," *Maarachot* 126 (1960), pp. 26–7, 48.

———. "Theory and Practice in the Military Profession," *Maarachot* 130 (August 1960), pp. 26–7, 53.

Bellamy, Chris. *The Future of Land Warfare* (New York: St. Martin's Press, 1987).

Ben-David, Alon. "Israel Introspective after Lebanon Offensive," *Jane's Defense Weekly*, Vol. 22 (August 2006), pp. 18–19.

Ben-Dor, Gabriel. "The Interface between the Military and the Academic World in Israel," paper presented at an international conference on 'The Decline of the Citizen Armies in Democratic States: Processes and Implications,' Bar-Ilan University, 18–19 June 2008.

Ben-Eliezer, Uri. "The Military and Civil Societies in Israel: Expressions of Anti-Militarism and Neo-Militarism in a Post-Hegemonic Era," in Majid Al Haj and Uri Ben-Eliezer (eds), *In the Name of Security: The Sociology of Peace and War in Israel in Changing Times* (Haifa: Haifa University, 2003) [Hebrew], pp.29–76.

———. *The Making of Israeli Militarism* (Bloomington: Indiana University Press, 1998).

Ben-Eliyahu, Eitan. "The Rear Will be the Decisive Factor," *NRG News*, 16 July 2006 <http://www.nrg.co.il/online/1/ART1/449/981.html>.

Ben-Gurion, David. *Behilahem Israel* (Tel Aviv: Am Oved, 1975) [Hebrew].

———. *Yichud Ve-Ye'ud* (Tel Aviv: Maarachot, 1971) [Hebrew].

Ben-Israel, Isaac. *The First Missile War* (Tel Aviv: The Security Studies Program, May 2007) [Hebrew].

———. "Security, Technology, and Future Battlefield," in Haggai Golan (ed.), *Israel's Security Web* (Tel Aviv: Maarachot, 2001) [Hebrew], pp. 269–327.

———. "The Military Buildup's Theory of Relativity," *Maarachot* 352–353 (August 1997), pp. 33–42.

———. "Philosophy and Methodology of Intelligence: The Logic of Estimate Process," *Intelligence and National Security*, Vol. 4, No. 4 (October 1989), pp. 660–718.

Ben-Moshe, Tuvia. "Liddell Hart and the Israel Defense Forces: A Reappraisal," *Journal of Contemporary History*, Vol. 16, No. 2 (April 1981), pp. 369–91

Benn, Aluf. "In Israel: Too Much to Leave to the Generals," *The Washington Post*, 18 August 2002.

Ben-Shalom, Uzi. "The Military Profession in Israel: The Case of the Tactical Command College," paper delivered at the Kinneret Institute, 14 December 2011.

Ben-Tzedef, Eviatar. "The Israel Defense Forces, 1996," *Outpost* (September 1996). <www.afsi.org/OUTPOST/96SEP/sep4.htm>.

Ben-Yishai, Ron. "Israel Air Force is Deadlier than Ever," *YNET*, 8 May 2014 <http://www.ynetnews.com/articles/0,7340,L-4517002,00.html>.

Bercuson, David J. "Up from the Ashes: The Re-Professionalization of the Canadian Forces After the Somalia Affair," in Cohen (ed.), *The New Citizen Armies*, pp. 159–69.

Berger, Thomas U. "Norms, Identity, and National Security in Germany and Japan," in Katzenstein (ed.), *The Culture of National Security*, pp. 317–56.

Bernstein, Nina. "The Strategists Fight a War about the War," *New York Times*, 6 April 2003.

Biddle, Stephen. "Military Power: A Reply," *Journal of Strategic Studies*, Vol. 28, No. 33 (June 2005), pp. 453–69.

———. *Military Power: Explaining Victory and Defeat in Modern Battle* (Princeton: Princeton University Press. 2004).

Blank, Stephen. "Soviet Forces in Afghanistan: Unlearning the Lessons of Vietnam," in Stephen Blank et al., *Responding to Low-Intensity Conflict Challenges* (Maxwell Air Base: Air University Press, 1990), pp. 53–176.

Blash, Edmund C. "Signal Forum: Network-Centric Warfare Pro & Con," <http://www.iwar.org.uk/rma/resources/ncw/ncw-forum.htm>.

Boëne, Bernard. "Trends in the Political Control of Post-Cold War Armed Forces," in Stuart Cohen (ed.), *Democratic Societies and Their Armed Forces: Israel in Comparative Context* (London: Frank Cass, 2000), pp. 73–88.

Bogner, Nahum. *Military Thought in the Hagahah* (Tel Aviv: MOD, 1998) [Hebrew].

Bond, Brian. *Liddell Hart: A Study of His Military Thought* (London: Cassell, 1977).

———. "Liddell Hart's Influence on Israeli Military Theory and Practice," *Journal of the Royal United Services Institute*, Vol. 121 (June 1974), pp. 83–9.

Bonen, Zeev. "Technology in War: Preliminary Lessons from the Gulf War," in JCSS Study Group, *War in the Gulf: Implications for Israel* (Tel Aviv: Jaffee Center for Strategic Studies, 1992), pp. 170–83.

Boot, Max. "The New American Way of War," *Foreign Affairs*, Vol. 82, No. 4 (July/August 2003), pp. 41–58.

Booth, Ken. *Strategy and Ethnocentrism* (New York: Holmes & Meier, 1979).

Brodie, Bernard. *Strategy in the Missile Age* (Princeton: Princeton University Press, 1959).

Brown, Michael E. (ed.), *The International Dimensions of Internal Conflict* (Cambridge: MIT Press, 1996).

———. "Introduction," in Brown (ed.), *The International Dimensions of Internal Conflict*, pp. 4–7.

Cameron, C.M. "The US Military's 'Two-Front War,' 1963–1988," in Farrell and Terriff (eds), *The Sources of Military Change*, pp. 119–38.

Carter, Ashton. "Responding to the Threats: Preventive Defense," paper presented at the conference on 'Challenges to Global and Middle East Security,' Jaffee Center for Strategic Studies and Belfer Center for Science and International Affairs, Herzliah, 15–16 June 1998.

Cassidy, Robert M. *Russia in Afghanistan and Chechnya: Military Strategic Culture and the Paradoxes of Asymmetric Conflict* (Carlisle: US Army War College, Strategic Studies Institute, 2003) <http://www.smallwarsjournal.com/documents/russia.pdf>.

Clausewitz, Carl von. *On War* (Princeton: Princeton University Press, 1976).

———. *The Principles of War* (Harrisburg: The Military Service Publishing Company, 1952).

Cohen, Eliot A. "The 'Major' Consequences of War," *Survival*, Vol. 41, No. 2 (Summer 1999), pp. 139–52.

———. "A Revolution in Warfare," *Foreign Affairs*, Vol. 75, No. 2 (March/April 1996), pp. 37–54.

———. "Technology and Supreme Command," *Foreign Affairs*, Vol. 75, No. 2 (March/April 1996), pp. 89–103.

———. *Citizens and Soldiers: The Dilemmas of Military Service* (Ithaca: Cornell University Press, 1985).

Cohen, Eliot. "An Intellectual Challenge," *Haaretz*, 20 September 1998.

Cohen, Eliot A. and John Gooch, *Military Misfortunes: The Anatomy of Failure in War* (New York: The Free Press, 1990).

Cohen, Eliot A., Michael J. Eisenstadt, and Andrew J. Bacevich. *Knives, Tanks, and Missiles: Israel's Security Revolution* (Washington, DC: Washington Institute for Near East Policy 1998).

Cohen, Eliot, Conrad Crane, Jan Horvath, and John Nagl. "Principles, Imperatives, and Paradoxes of Counterinsurgency," <http://usacac.army.mil/CAC/milreview/English/MarApr06/Cohen.pdf>.

Cohen, Stuart. "Light and shadows in US-Israel military ties, 1948–2010," *in* Robert Freedman (ed.), *Israel and the United States: Six Decades of US-Israeli Relations* (Boulder: Westview Press, 2012), pp. 143–64.

———. (ed.). *The New Citizen Armies* (London: Routledge, 2010).

———. (ed.), *Democratic Societies and Their Armed Forces: Israel in Comparative Context* (London: Frank Cass, 2000).

———. "Military Service in Israel: No Longer a Cohesive Force?" *Jewish Journal of Sociology*, Vol. 39 (1997), pp. 5–23.

———. "The Israel Defense Forces (IDF): From a People's Army to a Professional Military: Causes and Implications," *Armed Forces & Society*, Vol. 21, No. 2 (Winter 1995), pp. 237–54.

Conetta, Carl "The Wages of War: Iraqi Combatant and Noncombatant Fatalities in the 2003 Conflict," *Project on Defense Alternatives Research Monograph*, No. 8, 20 October 2003 <http://www.cbc.ca/news/iraq/issues_analysis/casualties_postiraqwar.html>.

Corbett, Julian S. *Some Principles of Maritime Strategy* (Annapolis: Naval Institute Press, 1988).

Cordesman, Anthony H. *Preliminary "Lessons" of the Israeli-Hezbollah War* (Washington DC: Center for Strategic and International Studies, 11 September 2006).

Cordesman, Anthony H., and Abraham Wagner. *The Lessons of Modern War, Vol. 2: The Iran-Iraq War* (Boulder: Westview, 1990).

Cornish, Paul David Livingstone, Dave Clemente and Claire Yorke. *Security and the UK's Critical National Infrastructure* (London: Chatham House, September 2011).

Corvisier, Andre (ed.), *Dictionary of Military History* (Oxford: Blackwell 1994).

David, Steven R. *Fatal Choices: Israel's Policy of Targeted Killing* (Ramat-Gan: BESA Center for Strategic Studies, 2002).

Davidson, Lance. "The Impact of Precision Guided Munitions on War," in Brassey's Yearbook 1984 (London: Brassey, 1984), pp. 237–52.

Dayan, Moshe. *The Vietnam Dairy* (Tel Aviv: Dvir, 1977) [Hebrew].

———. *Story of My Life* (Jerusalem: Edanim, 1976) [Hebrew].

De Atkine, Norwell. "Why Arabs Lose Wars?" *Middle East Quarterly*, Vol. 6, No, 4 (1999), pp. 17–27.

Demchak, Chris C. "Complexity and Theory of Networked Militaries," in Theo Farrell and Terry Terriff (eds), *The Sources of Military Change: Culture, Politics, Technology* (Boulder: Lynne Rienner, 2002), pp. 221–62.

———. "Technology's Burden, the RMA, and the IDF: Organizing the Hypertext Organization for Future 'Wars of Disruption'?" *Journal of Strategic Studies*, Vol. 24, No, 2 (June 2001), pp. 77–146.

Desch, Michael C. "Correspondence," *International Security*, Vol. 24, No. 1 (Summer 1999), pp. 156–80.

———. "Culture Clash: Assessing the importance of Ideas in Security Studies," *International Security*, Vol. 23, No. 1 (Summer 1998), pp. 141–70.

Douhet, Giulio. *The Command of the Air* (London: Faber & Faber, 1943),

Dror, Yehezkel. *A Grand Strategy for Israel* (Jerusalem: Academon, 1989) [Hebrew].

Duffield, John S. "Isms and Schisms: Culturalism versus Realism in Security Studies: Correspondence," *International Security*, Vol. 24, No. 1 (Summer 1999), pp. 172–80.

Earle, Edward M. (ed.). *Makers of Modern Strategy* (New Jersey: Princeton University Press, 1943).

Edwards, Sean J.A. *Swarming on the Battlefield: Past, Present, and Future* (Santa Monica: Rand, 2000).

Eilon, Giora. "If We Keep that Way, the State of Israel Might Collapse," Interview with Martin Van Creveld, *Al Hasharon*, 8 March 2002.

Eric M. Hammel, (1992). *Six Days in June: How Israel Won the 1967 Arab-Israeli War* (New York: Simon & Schuster, 1992).

Erickson, John. *The Soviet High Command* (London: St. Martin's Press, 1962).

Evans, Michael. "From the Long Peace to the Long War: Armed Conflict and Military Education and Training in the 21st Century," <http://www.defence.gov.au/jetwc/docs/publications%202010/PublcnsOccasional_310310_FromtheLongPeace.pdf>.

Even, Shmuel and David Siman-Tov. *Cyber Warfare: Concepts and Strategic Trends* (Tel Aviv: INSS, 2012).

Eytan, Raphael. *Story of a Soldier* (Tel-Aviv: Maariv, 1985) [Hebrew].

Farrell, Theo. "World Culture and Military Power," *Security Studies*, Vol. 14, No. 3 (July-September 2005), pp. 448–88.

———. "Isms and Schisms: Culturalism versus Realism in Security Studies: Correspondence," *International Security*, Vol. 24, No. 1 (Summer 1999), pp. 161–8.

———. "Culture and Military Power," *Review of International Studies*, Vol. 24, No. 3 (July 1998), pp. 407–16.

Farrell, Theo. and Terry Terriff (eds). *The Sources of Military Change: Culture, Politics, Technology* (Boulder: Lynne Rienner, 2002).

Feldman, Yotam. "Warhead," interview with Shimon Naveh, *Haaretz Supplement*, 26 October 2007.

Fikelstein, Menachem. "The Purity of Arms," <http://www.daat.ac.il/mishpat-ivri/skirot/235-2.htm>.

Finkel, Meir. *On Flexibility* (Tel Aviv: MOD, 2007) [Hebrew].

———. "The Cult of Technology in the IDF," *Maarachot* 407 (June 2006), pp. 40–5.

Finkel, Meir and Eitan Shamir. "From Whom Does the IDF Need to Learn?" *Maarachot* 433 (October 2010), pp. 28–35.

Fishman, Alex. "For Your Attention, Gabbi [Ashkenazi]," *Yediot Aharonot Weekend Supplement*, 26 January 2007.

Fishman, Alex. "Thanks to the Censorship," *Yediot Aharonot Weekend Supplement*, 11 May 2007.

———. "The Five-Day War," *Yediot Aharonot Weekend Supplement*, 17 April 2007.

———. "They Extinguished the Fire and Gained Time," *Yediot Aharonot Weekend Supplement*, 21 January 2005.

Ford, Peter. "Surveys Pointing to High Civilian Death Toll in Iraq," *Christian Science Monitor*, 22 May 2003.

Foster, Gregory D. "Research, Writing, and the Mind of the Strategist," *Joint Force Quarterly*, 11 (Spring 1996), pp. 111–15.

Fourth Generation Warfare, A special issue, *Contemporary Security Policy*, Vol. 26, No. 2 (August 2005), pp. 185–285.

Frederick II (the Great), The King of Prussia's Military Instruction to his Generals, Article VI, "Of the Coup D'Oeil," <http://www.kw.igs.net/~tacit/artofwar/frederick.htm#VI>.

Fuller, John F.C. *The Conduct of War 1789–1961* (New Brunswick, New Jersey: Rutgers University Press, 1961).

———. *Armament and History* (New York: Scribner, 1945).

Gaffney, H.H. *The American Way of War through 2020*, The CNA Corporation <http://www.au.af.mil/au/awc/awcgate/cia/nic2020/way_of_war.pdf>.

Gal, Reuven. *A Portrait of the Israeli Soldier* (Westport: Greenwood, 1986).

Gantzel, Klaus J. "War in the Post-World War II World: Some Empirical Trends and a Theoretical Approach", in David Turton (ed.), *War and Ethnicity: Global Connections and Local Violence* (San Marino: University of Rochester Press), pp. 125–38.

Gat, Azar. *The Origins of Military Thought* (Oxford: Clarendon Press, 1989).

———. *Policy and War in Modern Military Thought* (Tel Aviv: Maarachot, 1984) [Hebrew].

Gentry, John A. "Doomed to Fail: America's Blind Faith in Military Technology," *Parameters*, Vol. 22, No. 4 (Winter 2002–03), pp. 88–103.

Gibson, James W. *The Perfect War: Technowar in Vietnam* (New York: Atlantic Monthly Press, 1986).

Gilbert, Felix. "Machiavelli: The Renaissance of the Art of War," in Paret (ed.), *Makers of Modern Strategy*, pp. 11–31.

Gilboa, Eytan. "Educating Israeli officers in the Process of Peacemaking in the Middle East Conflict," *Journal of Peace Research*, Vol. 16, No. 2 (1979), pp. 155–62.

Glenn, Russell W. *All Glory Is Fleeting Insights from the Second Lebanon War* (Santa Monica: Rand, 2012).

Glick, Carolyn B. "An Interview with PM Ariel Sharon," *Jerusalem Post*, 26 September 2002.

Goerlitz, Walter. *History of the German General Staff 1657–1945* (New York: Praeger, 1957).

Golan, Haggai, and Shaul Shai (eds). *Limited Confrontation* (Tel Aviv: Maarachot, 2004) [Hebrew]

Golan, Haggai (ed.). *Israel's Security Web* (Tel-Aviv: Maarachot, 2001) [Hebrew].

Golani, Motti, *Wars Don't Just Happen* (Ben-Shemen: Modan, 2002) [Hebrew].

Gordon, Neve and George A. Lopez. *Terrorism in the Arab-Israeli Conflict*, in Valls (ed.), *Ethics in International Affairs*, pp. 99–113.

Gordon, Shmuel, *The Bow of Paris* (Tel Aviv: Poalim, 1997) [Hebrew].

———. "Winning from the Air is Still Possible," <http://www.ynet.co.il/articles/1,7340,L-3288914,00.html>.

Granot, Oded and Itzik Saban. Interview with Moshe Yaalon, *Maariv Weekend Supplement*, 16 February 2001.

Grau, Lester W. and Michael A. Gress. *The Soviet-Afghan War: How a Superpower Fought and Lost* (Kansas: University of Kansas Press, 2002).

Gray, Colin S. "Strategic Culture as Context: The First Generation of Theory Strikes Back," *Review of International Studies*, Vol. 25, No. 1 (January 1999), pp. 49–69.

Gray, Colin. *Irregular Enemies and the Defense of Strategy: Can the American Way of War Adapt?* (Carlisle: US Army College, March 2006).

———. *Modern Strategy* (Oxford: Oxford University Press, 1999).

———. *War, Peace and Victory* (New York: Simon & Schuster, 1990).

———. *Nuclear Strategy and National Style* (London: Hamilton, 1986).

———. "National Styles in Strategy: The American Example," *International Security*, Vol. 6, No. 2 (Fall 1981), pp. 21–47.

Gross, Emanuel. *The Struggle of Democracy against Terrorism*: Lessons from the United States, the United Kingdom, and Israel (Charlottesville: University of Virginia Press, 2006), pp. 220–46.

———. "Self-Defense against Terrorism: What Does It Mean? The Israeli Perspective," *Journal of Military Ethics*, Vol. 1, No. 2 (2002), pp. 91–108.

Gross, Michael L. "Assassination: Killing in the Shadow of Self-Defense," in J. Irwin (ed.), *War and Virtual War: The Challenge to Communities*, Amsterdam: Rodopi, 2004, pp. 99–116.

———. "Fighting by Other Means in the Mideast: A Critical Analysis of Israel's Assassination Policy," *Political Studies* Vol. 51, No, 2 (2003), pp. 350–68.

Grundman, Moshe (ed.), *Israel's Security 1967–1991: An Annotated Bibliography and Research Guide* (Tel-Aviv: Jaffee Center for Strategic Studies, 1992).

Guerlac, Henry. "Vauban: The Impact of Science on War," in Paret (ed.), *Makers of Modern Strategy*, pp. 64–90.

Guiora, Amos N. "Targeted Killing as Active Self-Defense," *Case Western Reserve Journal of International Law*, Vol. 36 (2004), pp. 319–34.

Gur, Mordechai. *Chief of the General Staff 1974–1978* (Tel-Aviv: Maarachot, 1998) [Hebrew].

———. "The IDF: Continuity versus Innovation," *Maarachot*, 261–262 (March 1978), pp. 4–6.

HaCohen, Gershon. "Educating Senior Officers," in *Is the IDF Prepared for Tomorrow's Challenges?* (Ramat Gan: BESA Colloquia on Strategy and Diplomacy 24, July 2008), pp. 33–4.

———. "Designing the Space and a Military Campaign during the Disengagement [from Gaza]," *Maarachot* 432 (August 2010), pp. 24–34.

———. "How Should the Curriculum in Military Colleges Look Like?" *Scrutinizing on National Security* 14 (September 2011).

Halutz, Dan. Lecture at the National Defense College, 28 January 2001.

Hammes, Thomas X. "War Evolves into the Fourth Generation," *Contemporary Security Policy*, Vol. 26, No. 2 (August 2005), pp. 254–63.

Hammond, Grant T. *The Mind of War: John Boyd and American Security* (Washington, Smithsonian Books, 2001).

Handel, Michael I., *Masters of War* (London: Frank Cass, 2001).

———. "Clausewitz in the Age of Technology," in Michael Handel (ed.), *Clausewitz and Modern Strategy* (London: Frank Cass, 1986), pp. 51–92.

Harel, Amos. "The Hundred Years War," *Alaxon*, 14 April 2013.

———. "A Flawed Operational Conception," *Haaretz*, 10 December 2006.

———. "Zero Tolerance," *Haaretz*, 7 April 2006.

Har'el, Dan. "The Change in Israel's Security," Lecture at the Interdisciplinary College, 17 March 2003 <http://idclawreview.org/2013/03/17/militraylimitations2012-pt1>.

Harkabi, Yehoshafat. *War and Strategy* (Tel Aviv: Maarachot, 1990) [Hebrew].

———. "From Guerilla Warfare to Guerilla War," in Yehoshafat Harkaby (ed.), *On Guerilla* (Tel Aviv: Maarachot, 1971 [Hebrew]), pp. 32–39.

Hecht, Eado. "Limited Confrontation: A Few General Features of a Unique Form of Warfare," in Golan and Shay (eds), *Low Intensity Conflict*, pp. 45–68.

Hersh, Seymour M. "Offense and Defense," *The New Yorker*, 7 April 2003.

Higham, Robin. *The Military Intellectuals in Britain: 1918–1939* (New Brunswick: Rutgers University Press. 1966).

Hirsch, Gal. "Urban Warfare," Military and Strategic Affairs, (April 2014), p. 25 <http://www.inss.org.il/uploadImages/systemFiles/HirschUrbanWarfare.pdf>.

———. "On Dinosaurs and Hornets: A Critical View on Operations Moulds in Asymmetric Conflicts," *RUSI Journal*, Vol. 148, No. 4 (August 2003), pp. 60–3.

Hisdai, Yaacov. "Ideologue versus Performer: IDF's Priest and Prophet," *Maarachot* 279–280 (May-June 1981), pp. 41–6.

Hoffman, Frank. "The Anatomy of the Long War's Failings," *FPRI's Newsletter*, Vol. 14 No. 16 (May 2009) <http://www.fpri.org/footnotes/1416.200905.hoffman.longwarsfailings.html#note5>.

———. "Preparing for Hybrid Wars," *Marine Corps Gazette*, Vol. 91, No. 3 (March 2007), pp. 57–61.

———. "Lessons From Lebanon: Hezbollah And Hybrid Wars," *The Evening Bulletin*, 5 September 2006 <http://www.theeveningbulletin.com/site/news.cfm?newsid=17152236&BRD=2737&PAG=461&dept_id=574088&rfi=6>.

———. "Complex Irregular Warfare: The Next Revolution in Military Affairs," *Orbis*, Vol. 50, No. 3 (Summer 2006), pp. 395–411.

———. "How Marines are Preparing for Hybrid Wars," *Armed Forces Journal* <http://www.armedforcesjournal.com/2006/03/1813952/ >.

Holborn, Hajo. "The Prusso-German School: Moltke and the Rise of the General Staff," in Paret (ed.), *Makers of Modern Strategy*, pp. 281–95.

Hopf, Ted. "The Promise of Constructivism in IR Theory," *International Security*, Vol. 23, No. 1 (Summer 1998), pp. 171–200.

Horowitz, Dan and Moshe Lissak. *Trouble in Utopia: The Overburdened Policy of Israel* (Tel Aviv: Am Oved, 1990) [Hebrew].

Howard, Michael. "The Forgotten Dimensions of Strategy," *Foreign Affairs*, Vol. 57, No. 5 (Summer 1979), pp. 975–86.

———. *War in European History* (London: Oxford University Press, 1976).

———. "Men against Fire: The Doctrine of the Offensive in 1914," in Paret (ed.), *Makers of Modern Strategy*, pp. 510–26.

Hughes, Daniel. *Moltke On the Art of War: Selected Writings* (New York: Presidio, 1995).

Huntington, Samuel P. *The Soldier and the State: The Theory and Politics of Civil-Military Relations* (Cambridge: Belknap Press, 1957).

IAF Chief General Benny Peled. Interview to the monthly *Nihul* – Israel's Management Magazine, November 1980.

Inbar, Efraim. *Rabin and Israel's National Security* (Baltimore: Johns Hopkins University Press, 1999), pp. 104–5.

———. *The Outlook on War of the Political Elite in the Eighties* (Jerusalem: The Leonard Davis Institute for International Relations, 1988).

Inbar, Efraim and Eitan Shamir. "Mowing the Grass," *Journal of Strategic Studies*," Vol. 37, No. 1 (February 2014), pp. 65–90.

Interview with General (retired) Uri Sagi, *Yediot Aharonot 7 Days Supplement*, 15 May 1998.

Interview with General Schwarzkopf, in "Hannibal and Desert Storm," *Timewatch* BBC Television, 1996.

Interview with IAF Chief General Shkedy, *Janes Defence Weekly*, Vol. 42, No. 1 (5 January 2005), p. 34.

Interview with Professor Avi Ravitzky, *Maariv* <http://www.nrg.co.il/online/archive/ART/138/748.html>.

Interview with Shelly Yechimovich, "Meet the Press" show, Israeli Television, Channel 2, 16 March 2002.

Jacobsen, Carl G. (ed.). *Strategic Power: USA/USSR* (New York: St. Martin's Press, 1990).

Janowitz, Morris. *The Professional Soldier: A Social and Political Portrait* (New York: The Free Press, 1960).

Jeffay, Nathan. "IDF Lawyers Now Coming Under Fire for Their Counsel during Gaza Conflict," *Forward*, 25 March 2009.

Joesten, Joachim. *The New Algeria* (Chicago: Follett, 1964).

Johnson, Paul. "Algeria War 1954–1962," <http://www.freerepublic.com/focus/f-news/672136/posts>.

———. *Intellectuals: From Marx and Tolstoy to Sartre and Chomsky* (London: Weidenfeld and Nicolson, 1988).

Johnston, Alastair Iain. "Cultural Realism and Strategy in Maoist China," in Katzenstein (ed.) *The Culture of National Security*, pp. 216–68.

———. "Thinking about Strategic Culture," *International Security*, Vol. 19, No. 4, (Spring 1995), pp. 32–64.

Jomini, Antoine-Henri. *Summary of the Art of War* (Urbana: University of Illinois Press, 1947).

Kaldor, Mary. *New and Old Wars: Organized Violence in a Global Era* (Oxford: Blackwell, 1999).

———. *New and Old Wars: Organized Violence in a Global Era* (Cambridge: Polity 1999).

Kasher, Asa. "The Morality of Preemptive Warfare," *Maariv*, 12 January 2001.

Kaspit, Ben. Interview with Ariel Sharon, *Maariv Weekend Supplement*, 13 April 2001.

Katzenstein, Peter (ed.). *The Culture of National Security: Norms and Identity in World Politics* (New York: Columbia University Press, 1996).

Keyfitz, Nathan, and Wilhelm Flieger. *World Population* (Chicago: Chicago University Press, 1968).

Kier, Elisabeth. *Imagining War: French and British Military Doctrine between the Wars* (Princeton: Princeton University Press, 1999).

———. "Culture and French Military Doctrine before World War II," in Katzenstein (ed.), *The Culture of National Security*, pp. 186–215.

Kimmerling, Baruch. "Patterns of Militarism in Israel," *European Journal of Sociology*, Vol. 34, No. 2 (1993), pp. 115–33.

Klein, Yitzhak. "A Theory of Strategic Culture," *Comparative Strategy*, Vol. 10, No. 1 (January-March 1991), pp. 3–23.

Klieman, Aaron S. "Lavi: The Lion Has Yet to Roar," *Journal of Defense and Diplomacy*, Vol. 4 (August 1986), pp. 22–9.

Knott, Steven W. *Knowledge Must Become Capability: Institutional Intellectualism as an Agent for Military Transformation* (Carlisle Barracks: US Army War College, 2004).

Kober, Avi. "From Heroic to Post-Heroic Warfare: Israel's Way of War in Asymmetrical Conflicts," *Armed Forces & Society*, Vol. 41 (January 2015), pp. 96–122. First published online on 1 August 2013.

———. "Can the IDF Afford a Small Army? *BESA Perspectives*," No. 209 (July 18, 2013).

———. "Iron Dome: Has the Euphoria Been Justified?" *BESA Perspectives*, No. 199 (25 February 2013).

———. "What Happened to Israeli Military Thought?" *Journal of Strategic Studies*, Vol. 34, No. 5 (October 2011), pp. 707–32.

———. "The Rise and Fall of Israeli Operational Art," in Martin van Creveld and John A. Olsen (eds.), *Operational Art: From Napoleon to the Present* (New York: Oxford University Press, 2011), pp. 166–94.

———. *Israel's Wars of Attrition* (New York: Routledge, 2009).

———. "Operational and Technological Incentives and Disincentives for Force Transformation," in Stuart Cohen (ed.), *The New Citizen Armies* (London: Routledge, 2009), pp. 77–91.

———. "The IDF in the Second Lebanon War: Why the Poor Performance?" *Journal of Strategic Studies*, Vol. 31, No. 1 (February 2008), pp. 3–40.

———. "Israeli Military Thinking as Reflected in *Maarachot* Articles, 1948–2000," *Armed Forces & Society*, Vol. 30, No. 1 (Fall 2003), pp. 141–60.

———. "Low-Intensity Conflicts: Why the Gap between Theory and Practise?" *Defense & Security Analysis*, Vol. 18, No, 1 (March 2002), pp. 15–38.

Kopp, Carlo. "Understanding Network Centric Warfare," Australian Aviation (January/ February 2005) <http://www.ausairpower.net/TE-NCW-JanFeb-05.html>.

Koren-Lieber, Stella. "A Small but High-Tech Military," *Globes*, 29 August 1999.

Krauthammer, Charles. "Gulf War II is First of its Kind," *TownHall.com*, 10 April 2003.

Kremnitzer, Mordechai. *Are All Actions Acceptable in the Face of Terror? On Israel's Policy of Preventive (Targeted) Killing in Judea, Samaria and Gaza* (Jerusalem: Israel Democracy Institute, 2005) [Hebrew].

Kretzmer, David. "Targeted Killing of Suspected Terrorists," in *The European Journal of International Law*, Vol. 16, No. 2 (2005), pp. 171–212.

Lam, Amira. "We Betrayed our Constituency," *Yediot Aharonot Weekend Supplement*, 1 September 2006.

———. "The General Who Knew When to Retire," *Yediot Aharonot 7 Days Supplement*, 17 July 1998.

Lambeth, Benjamin S. *The Transformation of American Air Power* (Ithaca: Cornell University Press, 2000).

Lanir, Zvi. "Who Needs the Concept Jointness," *Maarachot* 401 (June 2005), pp. 20–27.

———. *War by Choice* (Tel Aviv: Jaffee Center for Strategic Studies, 1985) [Hebrew].

Lappin, Yaakov. "Israel vs. the Iran-Hizballah Axis," *BESA Center Perspectives Paper*, No. 221 (14 November 2013) <http://besacenter.org/perspectives-papers/israel-vs-iran-hizballah-axis/>.

Laqueur, Walter. *The New Terrorism, Fanaticism and the Arms of Mass Destruction* (New York: Oxford University Press, 1999).

Laskov, Haim and Meir Zorea. "Should One Wage War," *Maariv*, 10 October 1965.

Laughbaum, R. Kent. *Synchronizing Airpower and Firepower in the Deep Battle* (Maxwell Air Force Base: Air University Press, January 1999) <http://www.dtic.mil/cgi-bin/GetTRDoc?AD=ADA392508>.

Lawrence, T.E. *Revolt in the Desert* (New York: G.H. Doran, 1927).

———. "Science of Guerilla Warfare," <http://www.bellum.nu/literature/lawrence001.html>.

Legro, Jeffrey W. *Cooperation under Fire: Anglo-German Restraint during World War II* (Ithaca: Cornell University Press, 1995).

———. "Military Culture and Inadvertent Escalation in World War II," *International Security*, Vol. 18, No. 4 (Spring 1994), pp. 108–42.

Lendon, J.E. *Soldiers and Ghosts: A History of Battle in Classical Antiquity* (New Haven: Yale University Press, 2005).

Levenberg, Haim. *Military Preparations of the Arab Community in Palestine: 1945–1948* (London: Routledge, 1993).

Levite, Ariel. *Offense and Defense in Israeli Military Doctrine* (Tel Aviv: Jaffee Center for Strategic Studies, 1988).

Levy, Yagil. "From the Citizen Army to the Market Army," in Cohen (ed.), *The New Citizen Armies*, pp. 196–214.

———. *Israel's Materialist Militarism* (Lanham: Lexington Books, 2007).

———. "Materialist Militarism," *Alternative Information Center*, 8 August 2006 <http://www.alternativenews.org/news/english/materialist-militarism-20060808.html>.

———. "The War of the Peripheries: A Social Mapping of IDF Casualties in the Al-Aqsa Intifada," *Social Identities*, Vol. 12, No. 3 (May 2006), pp. 309–24.

Levy-Barzilai, Vered. "The High and the Mighty," *Haaretz*, 21 August 2002.

Libel, Tamir. "IDF Operational Level Doctrine and Education During the 1990s," *Defense and Security Analysis*, Vol. 26, No. 3 (September 2010), pp. 321–4.

Liddell Hart, B.H. *Strategy* (London: Faber & Faber, 1967).

———. *Thoughts on War* (London: Faber & Faber, 1943).

Liddell Hart, Basil H. *Why Don't We Learn from History?* (London: Allen & Unwin, 1972).

Lifshitz, Yaacov. "Managing Security Resources in the 2000s," in Golan (ed.), *Israel's Security Web*, pp. 42–70.

Linn, Brian M. *The Echo of Battle: The Army's Way of War* (Cambridge: Harvard University Press, 2007).

———. "The American Way of War' Revisited," *The Journal of Military History*, Vol. 66, No. 2 (April 2002), pp. 501–33.

Loeb, Vernon. "Rumsfeld Faulted for Troop Dilution," *Washington Post*, 30 March 2003.

Lord, Amnon. "From the Chinese Farm to the Ministry of Defense," <http://rotter.net/cgi-bin/forum/dcboard.cgi?az=show_thread&forum=gil&om=5775&omm=243>.

Lord, Amnon. "The Air Went Out," *Makor Rishon*, 2 November 2006.

Luft, Gal. "The Logic of Israel's Targeted Killing," *The Middle East Quarterly*, Vol. 10 (2003) <http://www.meforum.org/article/515>.

Luttwak, Edward N. "Misreading the Lebanon War," *Jerusalem Post*, 21 August 2006.

———. "Post-Heroic War," *Maarachot* 374–375 (February 2001), pp. 4–9.

———. "A Post-Heroic Military Policy," *Foreign Affairs*, Vol. 75, No. 4 (July/Aug. 1996), pp. 33–44.

———. "Toward Post-Heroic Warfare," *Foreign Affairs*, Vol. 74, No. 3 (May/June 1995), pp. 109–2.

———. "Where Are the Great Powers?" *Foreign Affairs*, Vol. 73, No. 4 (July/Aug. 1994), pp. 23–8.

———. "On the Need to Reform American Strategy," in Philip S. Kronenberg (ed.), *Planning US Security: Defense Policy in the Eighties* (New York: Pergamon, 1982), pp. 13–29.

Luttwak, Edward and Dan Horowitz. *The Israeli Army* (London: Allen Lane, 1975).

Lynn, John A. *Battle: A History of Combat and Culture* (Boulder: Westview Press, 2003).

Machiavelli, Niccolo. *The Art of War* (Indianapolis: Bobbs Merrill, 1965).

MacIsaac, David. "Voices from the Central Blue: The Air Power Theorists," in Paret (ed.), *Makers of Modern Strategy*, pp. 624–47.

Mahan, Alfred Thayer. *The Influence of Sea Power Upon History, 1660–1783* (Boston: Little, Brown, 1940).

Maoz, Zeev. *Defending the Holy Land* (Ann Arbor: University of Michigan Press, 2006).

Marrero, Abe F. "The Tactics of Operation Cast Lead," in Scott C. Farquhar, *Back to Basics: Study of the Second Lebanon War and Operation Cast Lead* (Fort Leavenworth, Kansas: US Army Combined Arms Center, May 2009), pp.83–102;

Marx, Karl and Friedrich Engels. *Die Klassenkampfe in Frankreich, 1848–1850* (Berlin: Vorwärts, 1895).

Mason, R.A. "Innovation and the Military Mind," <http://www.au.af.mil/au/awc/awcgate/au24-196.htm>.

Mathers, Jennifer G. "Reform and the Russian Military," in Farrell and Terriff (eds), *The Sources of Military Change*, pp. 161–84.

Matthews, Lloyd J. "The Uniformed Intellectual and His Place in American Arms, Part I," *Army*, Vol. 52, No. 7 (July 2002), pp. 18–20.

———. "The Uniformed Intellectual and His Place in American Arms, Part II," *Army*, Vol. 52, No. 8 (August 2002), pp. 31–40.

Mattis, James N. and Frank Hoffman, "Future Warfare: The Rise in Hybrid Wars," *Proceedings*, Vol. 132, No. 11 (November 2005), pp. 18–19.

Mattis, James N. "Assessment of Effects Based Operations," <http://smallwarsjournal.com/documents/usjfcomebomemo.pdf>.

McGhie, Ian A. "Cyber-Warfare: Vital Ground, 'Emperor's New Clothes,' or Strategic Paralysis?" Dissertation published by the Royal College of Defense Studies, July 2012

<http://www.da.mod.uk/colleges/rcds/publications/seaford-house-papers/2012-seaford-house-papers/shp-2012-mcghie.pdf/view>.

Mearsheimer, John J. *Liddell Hart and the Weight of History* (Ithaca: Cornell University Press, 1988).

Mem, Benny. "The Peace for Galilee War: Main Operations," *Maarachot* 284 (September 1982), pp. 24–48.

Merton. Robert. *Social Theory and Social Structure* (Illinois: The Free Press, 1949).

Metz, Steven. *Armed Conflict in the 21st Century: The Information Revolution and Post-Modern Warfare* (Carlisle: US Army War College, April 2000).

Michael, Kobi. "Military Knowledge and Weak Civilian Control in the Reality of Low Intensity Conflict: The Israeli Case," *Israel Studies*, Vol. 12, No. 1 (Spring 2007), pp. 28–52.

———. "The Dilemma behind the Classical Dilemma of Civil-Military Relations," *Armed Forces & Society*, Vol. 33, No. 4 (July 2007), pp. 518–46.

Milstein, Uri. television interview (5 February 2012) <https://www.youtube.com/watch?v=hZZc4pKdLaE>.

———. "The IDF's March of Folly," *Nativ*, Vol. 116 (May-June 2007), pp. 46-54.

———. *Krissa Velik'ha* [Collapse and its Lessons] (Tel Aviv: Yediot Aharonot, 1993) [Hebrew].

Moliere. *The Middle Class Gentleman* (*Le Bourgeois Gentilhomme*), Part 1 <http://www.fullbooks.com/The-Middle-Class-Gentleman-Le-Bourgeois1.html>.

Morris, Benny. *Righteous Victims* (Tel Aviv: Am Oved, 2003) [Hebrew].

———. *Israel's Border Wars 1949–1956* (Tel-Aviv: Am Oved, 1996) [Hebrew].

Moskos, Charles C. "Institutional and Occupational Trends in Armed Forces: An Update," *Armed Forces & Society*, Vol. 12, No. 3 (1986), pp. 377–82.

Moskos, Charles C. et. al., *The Postmodern Military* (New York: Oxford University Press, 2000).

———. "Armed Forces after the Cold War," in Moskos et. al., *The Postmodern Military*, pp. 1–13.

Mula, Shosh and Assefa Peled. "Testimonies from the Heart," *Yediot Aharonot Weekend Supplement*, 17 November 2006.

Mumford, Andrew and Bruno Reis. "Constructing and Deconstructing Warrior-Scholars," in Andrew Mumford and Bruno Reis (eds), *The Theory and Practice of Counter-Insurgency: Warrior-Scholars in Irregular War* (New York: Routledge, 2013), pp. 4–17.

Murray, Williamson. "The Army's Advanced Strategic Art Program," *Parameters*, Vol. 30 (Winter 2000–01), pp. 31–9.

———. "Military Culture Does Matter," *FPRI Wire*, Vol. 7, No. 2 (January 1999) <http://www.fpri.org/fpriwire/0702.199901.murray.militaryculturedoesmatter.html>.

Nachmani, Amikam. "Generals at Bay in Post-War Palestine," *Journal of Strategic Studies*, Vol. 6, No. 4 (December 1983), pp. 66–83.

Naveh, Shimon. *Operational Art and the IDF: A Critical Study of a Command Culture* (Washington, DC: Center for Strategic and Budgetary Assessment, 2007).

Nelsen II, John T. "Auftragstaktik: A Case for Decentralized Battle," *Parameters*, Vol. 17 (September 1987), pp. 22–7.

Nietzsche, Friedrich. *Human, All Too Human* (Cambridge: Cambridge University Press, 1996).

Nir, Shmuel. "The Nature of the Limited Confrontation," in Golan and Shai (eds), *Low-Intensity Conflict*, pp. 19–44.

O'Hanlon, Michael. *Technological Change and the Future of War* (Washington, DC: Brookings Institution, 2000).

O'Ballance, Edgar. *The Algerian Insurrection 1954–1962* (London: Faber & Faber, 1967).

O'Sullivan, Arieh. "What a Riot," *Jerusalem Post*, 4 January 2004.

Olmert, Ehud. Speech before the Knesset, 17 July 2006.

Olson, William J. "Preface: Small Wars Considered," *Annals of the American Academy of Political and Social Studies*, Vol. 541 (September 1995), pp. 8–19.

Opall-Rome, Barbara. Interview with Major-General Benjamin Gantz, Chief of the IDF's Army, *Defense News*, 28 August 2006.

Oren, Amir. "Afeka, not Ofakim," *Haaretz*, 17 September 2010.

Ott, Steven J. *The Organizational Culture Perspective* (Pacific Grove: Brooks/Cole, 1989).

Palmer, R.R. "Frederick the Great, Guibert, Bülow: From Dynastic to National War," in Paret (ed.), *Makers of Modern Strategy*, pp. 91–119.

Paret, Peter (ed.). *Makers of Modern Strategy from Machiavelli to the Nuclear Age* (Princeton: Princeton University Press, 1986).

———. *French Revolutionary Warfare from Indochina to Algeria: the Analysis of a Political and Military Doctrine* (London: Pall Mall Press, 1964).

Pedatzur, Reuven. *The Arrow System* (Tel-Aviv: Jaffee Center for Strategic Studies 1993) [Hebrew].

———. "F-16s Would Cost Israel Half the Price of Lavi," *Jane's Defence Weekly*, 9 May 1987, p. 865.

Peres, Shimon. "Upgrading War, Privatizing Peace," <http://www.haaretz.com/hasen/spages/756832.html>.

———. *The Next Phase* (Tel Aviv: Am Hasseffer, 1965) [Hebrew].

Peri, Yoram. *Generals in the Cabinet Room* (Washington, DC: United States Institute of Peace Press, 2006).

Pfeffer, Anshel. "IDF to Seek Legal Advice during Future Conflicts," *Haaretz*, 6 January 2010.

Phillips, Thomas R. (ed.). *Roots of Strategy* (Harrisburg: The Military Service Publishing Company, 1940).

Picq, Ardant du. *Battle Studies* (Harrisburg, Pennsylvania: The Military Service Publishing Company, 1946).

Plato. *The Republic* (London: Methuen & Co., 1906).

Pollack, Kenneth M. *Arabs at War: Military Effectiveness 1948–1991* (Lincoln: University of Nebraska Press, 2002).

Porch, Douglas. "Military 'Culture' and the Fall of France in 1940: A Review Essay," *International Security*, Vol. 24, No. 4 (Spring 2000), pp. 157–80.

Posen, Barry. *The Sources of Military Doctrine: France, Britain, and Germany Between the World Wars* (Ithaca: Cornell University Press, 1984).

Postman, Niel. *Technopoly: The Surrender of Culture to Technology* (New York: Vintage Books, 1993).

Quinlivan, James. "Burden of Victory: The Painful Arithmetic of Stability Operations," *RAND Review*, Vol. 27, No. 3 (Summer 2003) <http://rand.org/pubs/corporate_pubs/2007/RAND_CP22–2003–08.pdf>.

———. "Force Requirements in Stability Operations," *Parameters*, Vol. 23 (Winter 1995), pp. 59–69.

Rabin, Yitzhak. *Pinkas Sharut* (Tel-Aviv: Maariv, 1979) [Hebrew].

Rapaport, Amir. "Back to the Traditional, Vernacular Language," *NRG*, 12 October 2008. <http://www.nrg.co.il/online/1/ART1/798/019.html>.

———. *Friendly Fire* (Tel Aviv: Maariv, 2007) [Hebrew].

Ravid, Barak. "Olmert: In This War There is No Once and For All.," <http://www.nrg.co.il/online/1/ART1/506/249.html>.

Reuters and Anshel Pfeffer, "How Cyber Warfare Has Made MI a Combat Arm of the IDF," *Haaretz*, 16 December 2009.

Rice, Condoleezza. "The Making of Soviet Strategy," in Paret (ed.), *Makers of Modern Strategy*, pp. 666–8.

Ringel-Hoffman, Ariella. "This is not how a War Should be Conducted," *Yediot Aharonot Weekend Supplement*, 23 March 2007.

Ritter, Gerhard. *The Sword and the Scepter: The Problem of Militarism in Germany* (Coral Cables: University of Miami Press, 1973).

Rosen, Stephen P. *Winning the Next War* (Ithaca: Cornell University Press, 1991).

Rosinski, Herbert. *The German Army* (New York: Praeger 1966).

Roszak, Theodore. *The Cult of Information: A Neo-Luddite Treatise on High-Tech, Artificial Intelligence, and the True Art of Thinking* (Berkeley: University of California Press, 1994).

Rotem, Avraham. "Is a Small and Smart Military a Vision or a Legend?" in Golan (ed.), *Israel's Security Web*, pp. 71–93.

Rothenberg, Gunther E. "Maurice of Nassau, Gustavus Adolphus, Raimondo Montecuccoli and the Military Revolution of the Seventeenth Century," in Paret (ed.), *Makers of Modern Strategy*, pp. 38–63.

Rothschild, J. "Culture and War," in Stephanie G. Neuman and Robert E. Harkavy (eds), *The Lessons of Recent Wars in the Third World, Vol. 2* (Lexington: Lexington Books, 1987), pp. 53–72.

Rowen, Henry S. *Intelligent Weapons: Implications for Offense and Defense* (Tel-Aviv: Jaffee Center for Strategic Studies, 1988).

Rubin, Uzi. "Iron Dome: A Dress Rehearsal for War?" *BESA Perspectives*, No. 173 (July 2012).

———. "Iron Dome in Action: A Preliminary Evaluation," BESA Perspectives, No. 151 (24 October 2011).

Samson, Elizabeth. "Warfare through Misuse of International Law," *BESA Perspectives*, No. 73 (23 March 2009).

Saxe, Marshal Maurice de. "My Reveries Upon the Art of War," in Phillips (ed.), *Roots of Strategy*, pp. 249–55.

Schalk, David. *War and the Ivory Tower* (New York: Oxford University Press, 1991).

Schiff, Zeev. "The Foresight Saga," <http://www.haaretz.com/hasen/spages/749268.html>.

Schoenberger, Karl. "Civilian Casualties will Erode Support, Inflame Hatred among Iraqis," *San Jose Mercury News*, 26 March 2003.

Schön, Donald A. *The Reflective Practitioner: How Professionals Think in Action* (London: Temple Smith, 1983).

Schueftan, Dan. "Beyond the Relative Advantage," *Maarachot* 356–357 (March 1998), pp. 70–79.

Seversky, Alexander P. de. *Air Power* (London: H. Jenkins, 1952).

Shabtay, Shay. "The Campaign between the Wars," *Maarachot* 445 (October 2011), pp. 24–7.

Shamir, Eitan. "A Very Sharp Eye: Moshe Dayan's Counterinsurgency Legacy in Israel," in Andrew Mumford and Bruno C. Reis (eds), *The Theory and Practice of Irregular Warfare: Warrior-Scholarship in Counter-Insurgency* (London: Routledge, 2013), pp. 84–104.

———. *Transforming Command: The Pursuit of Mission Command in the US, British, and Israeli Armies* (Stanford: Stanford University Press, 2011).

———. "When did a Big Mac Become Better than a Falafel? The Americanization Process of the IDF (1973–2006)," Paper Presented at the ISA Annual Conference, Montreal, March 2011.

Shamir, Moshe. "On Changes in the Inter-Arm Command and Control Training," *Maarachot* 396 (September 2004), pp. 20–25.

Shapir, Yiftah S. "Non-Conventional Solutions for Non-Conventional Dilemmas?" *Journal of Strategic Studies*, Vol. 24, No. 2 (June 2001), pp. 153–63.

Shapira, Anita. *Yigal Allon, Native Son: A Biography* (Philadelphia: University of Pennsylvania Press, 2007).

Sharet, Moshe. Speech before MAPAI party members on 22 May 1954, The Labor Party archives, Beit Berl.

Sharon, Ariel. Speech to the Nation, Israeli Television, Channel 1, 2 April 2002.

———. (with David Chanoff). *Warrior* (Tel-Aviv: Steimatzky, 1989).

Shavit, Ari. Interview with General (retired) Yossi Peled, *Haaretz Weekend Supplement*, 20 October 2006.

Shavit, Yaacov. *Self-Restraint or Reaction* (Ramat Gan: Bar-Ilan University Press, 1983) [Hebrew].

Shaw, George Bernard. *The Devil's Disciple* (Baltimore: Penguin Books, 1955).

Shelah, Ofer and Yoav Limor. *Captives in Lebanon* (Tel Aviv: Yediot Aharonot, 2007) [Hebrew].

Shelah, Ofer. *The Israeli Army: A Radical Proposal* (Or Yehuda: Kinneret, Zmora-Bitan, Dvir, 2003) [Hebrew].

Shy, John. "Jomini," in Paret (ed.), *Makers of Modern Strategy*, pp. 148–52.

Simpkin, Richard. *Deep Battle: The Brainchild of Marshal Tukhachevsky* (London: Brassey's, 1987).

Sivard, Ruth L. (ed.). *World Military and Social Expenditures* (Washington DC: World Priorities, 1987), pp. 29–31.

Smith, Edward A. *Effects Based Operations: Applying Network-Centric Warfare in Peace, Crisis, and War* (Washington, DC: Department of Defense Command and Control Research Program, 2002).

Snider, Don M. and Gayle Watkins (eds.). *The Future of the Army Profession* (New York: McGraw Hill, 2003).

Sofaer, Abraham D. "Response to Terrorism: Targeted Killing is a Necessary Option," *San Francisco Chronicle*, 26 March 2004 <http://www.sfgate.com/cgi-bin/article.cgi?file=/chronicle/archive/2004/03/26/EDGK65QPC41.DTL>.

Sondhaus, Lawrence. *Strategic Culture and Ways of War* (London: Routledge, 2006).

Stanley-Mitchell, Elizabeth A. "The Digital Battlefield: What Army Transformation Efforts Say about its Professional Jurisdiction," in Snider and Watkins (eds), *The Future of the Army Profession*, pp. 155–78.

Statman, Daniel. "Targeted Killing," *Theoretical Inquiries in Law*, Vol. 5, No. 1 (January 2004), pp. 179–98.

Stav, Arieh and Baruch Koroth (eds). *Ballistic Missiles: The Threat and Response* (Tel-Aviv: Yediot Aharonot, 1999) [Hebrew].

Steinberg, Gerald M. "Lessons of the Lavi," *Midstream*, Vol. 33 (November 1987), pp. 3–6.

Sullivan, Gordon R. and James M. Dubik. "War in the Information Age," *Military Review*, Vol. 74, No. 4 (April 1994), pp. 46–62.

Tal, Israel. *National Security: The Few against the Many* (Tel-Aviv: Dvir, 1996) [Hebrew].

———. "Offense and Defense in the Wars of Israel," *Maarachot* 311 (March 1988), pp. 4–7.

———. "The Tank at Present and in the Future," *Maarachot* 281 (November 1981), pp. 2–7.

———. "Introduction," in Douglass Orgill, *The Tank: Studies in the Development and Use of a Weapon* (Tel Aviv: Maarachot, 1980) [Hebrew], pp. 7–9.

———. "Israel's Security Doctrine," *Maarachot* 253 (December 1976), pp. 2–9,

Tamari, Dov. "Is the IDF Capable of Changing in the Wake of the Second Lebanon War?" *Maarachot* 415 (November 2007), pp. 26–41.

———. "Offense or Defense: Do We Have a Choice?" *Maarachot* 289–290 (October 1983), pp. 5–11.

———. "Thoughts on Tactics," *Maarachot* 273–4 (May-June 1980), pp. 2–5.

Tamari, Dov, and Meir Klifi. "The IDF's Operational Conception," *Maarachot* 423 (February 2009), pp. 26–41.

Terriff, Terry and Theo Farrell. "Military Change in the New Millennium," in Farrell and Terriff (eds), *The Sources of Military Change*, pp. 270–1.

Testimony by General (retired) Amatzia Chen, a former instructor at the National Defense College <http://www.kav.org.il/100994/647>.

The Limited Confrontation (Tel Aviv: IDF Training and Doctrine Department, 2001 [Hebrew].

The State Comptroller's Annual Report 57A, December 2006 <http://www.mevaker.gov.il/serve/contentTree.asp?bookid=474&id=57&contentid=&parentcid=undefined&sw=1024&hw=698>.

The State's Comptroller's Annual Report No. 37 for 1986.

The US Government Counterinsurgency Guide January 2009 <http://www.state.gov/documents/organization/119629.pdf>.

The Winograd Commission's final report <http://www.vaadatwino.org.il/pdf/דוח%20סופי.pdf>.

The Winograd Commission's Interim Report <http://www.vaadatwino.org.il/pdf/מאוחד%20סופי%20לאינטרנט.pdf>.

Thomas, Ward. "The New Age of Assassination," *SAIS Review*, Vol. 25, No. 1 (Winter–Spring 2005), pp. 27–39.

Thompson, Loren B. (ed.). *Low-Intensity Conflict: The Pattern of Warfare in the Modern World* (Lexington: Lexington Books, 1989).

Thompson, Loren B. "Low-Intensity Conflict: An Overview", in Thompson (ed.), *Low-Intensity Conflict*, pp. 1–23.

Tolstoy, Leo. *War and Peace* (New York: Vintage, 2008).

Turton, David. "Introduction", in David Turton (ed.), *War and Ethnicity: Global Connections and Local Violence* (San Marino: University of Rochester Press, 1997), pp. 125–38.

US Joint Forces Command, *Effects-based Operations White Paper Version 1.0* (Norfolk, VA: Concepts Department J9, 2001).

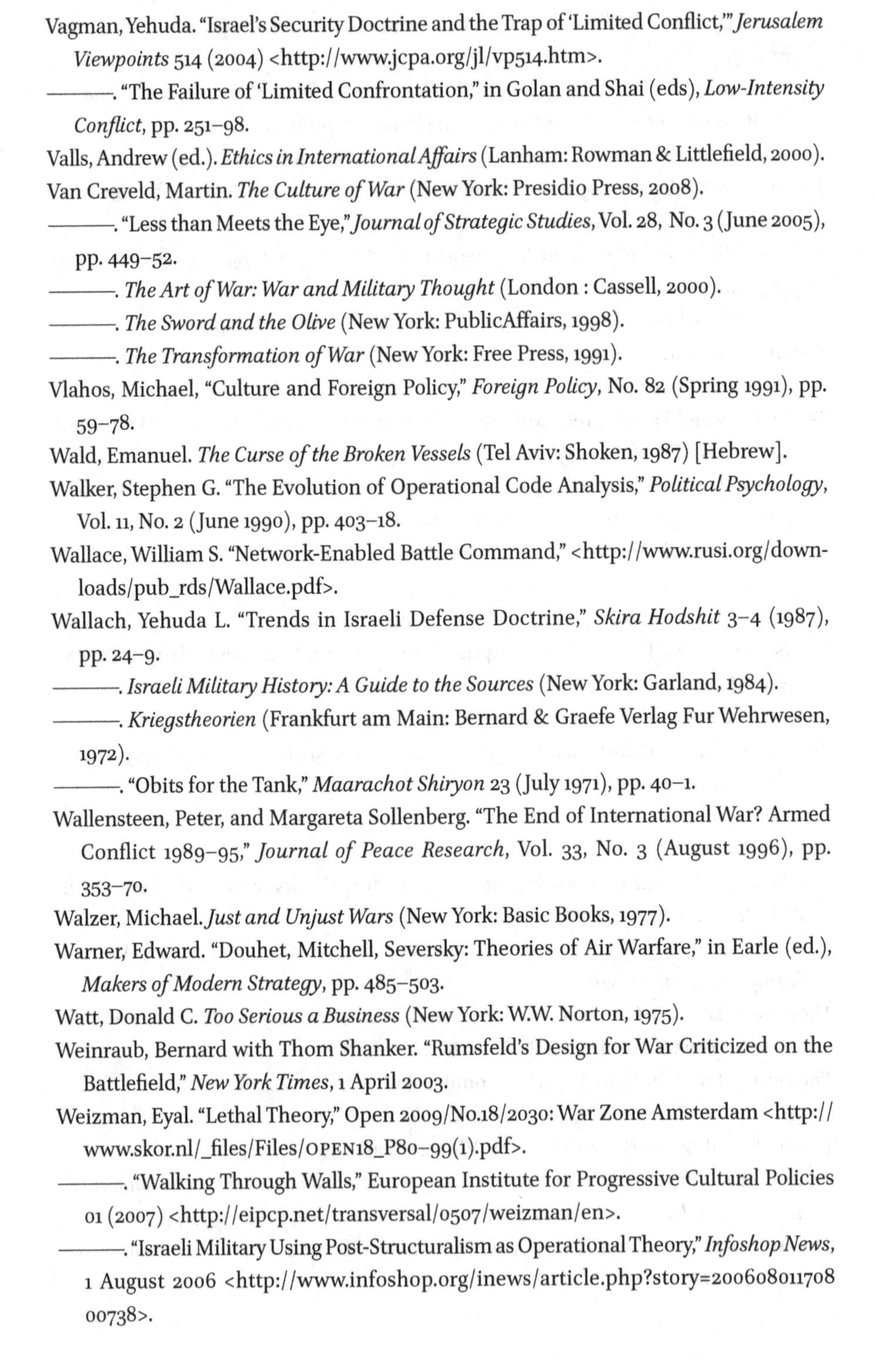

Vagman, Yehuda. "Israel's Security Doctrine and the Trap of 'Limited Conflict,'" *Jerusalem Viewpoints* 514 (2004) <http://www.jcpa.org/jl/vp514.htm>.

———. "The Failure of 'Limited Confrontation,'" in Golan and Shai (eds), *Low-Intensity Conflict*, pp. 251–98.

Valls, Andrew (ed.). *Ethics in International Affairs* (Lanham: Rowman & Littlefield, 2000).

Van Creveld, Martin. *The Culture of War* (New York: Presidio Press, 2008).

———. "Less than Meets the Eye," *Journal of Strategic Studies*, Vol. 28, No. 3 (June 2005), pp. 449–52.

———. *The Art of War: War and Military Thought* (London : Cassell, 2000).

———. *The Sword and the Olive* (New York: PublicAffairs, 1998).

———. *The Transformation of War* (New York: Free Press, 1991).

Vlahos, Michael, "Culture and Foreign Policy," *Foreign Policy*, No. 82 (Spring 1991), pp. 59–78.

Wald, Emanuel. *The Curse of the Broken Vessels* (Tel Aviv: Shoken, 1987) [Hebrew].

Walker, Stephen G. "The Evolution of Operational Code Analysis," *Political Psychology*, Vol. 11, No. 2 (June 1990), pp. 403–18.

Wallace, William S. "Network-Enabled Battle Command," <http://www.rusi.org/downloads/pub_rds/Wallace.pdf>.

Wallach, Yehuda L. "Trends in Israeli Defense Doctrine," *Skira Hodshit* 3–4 (1987), pp. 24–9.

———. *Israeli Military History: A Guide to the Sources* (New York: Garland, 1984).

———. *Kriegstheorien* (Frankfurt am Main: Bernard & Graefe Verlag Fur Wehrwesen, 1972).

———. "Obits for the Tank," *Maarachot Shiryon* 23 (July 1971), pp. 40–1.

Wallensteen, Peter, and Margareta Sollenberg. "The End of International War? Armed Conflict 1989–95," *Journal of Peace Research*, Vol. 33, No. 3 (August 1996), pp. 353–70.

Walzer, Michael. *Just and Unjust Wars* (New York: Basic Books, 1977).

Warner, Edward. "Douhet, Mitchell, Seversky: Theories of Air Warfare," in Earle (ed.), *Makers of Modern Strategy*, pp. 485–503.

Watt, Donald C. *Too Serious a Business* (New York: W.W. Norton, 1975).

Weinraub, Bernard with Thom Shanker. "Rumsfeld's Design for War Criticized on the Battlefield," *New York Times*, 1 April 2003.

Weizman, Eyal. "Lethal Theory," Open 2009/No.18/2030: War Zone Amsterdam <http://www.skor.nl/_files/Files/OPEN18_P80–99(1).pdf>.

———. "Walking Through Walls," European Institute for Progressive Cultural Policies 01 (2007) <http://eipcp.net/transversal/0507/weizman/en>.

———. "Israeli Military Using Post-Structuralism as Operational Theory," *Infoshop News*, 1 August 2006 <http://www.infoshop.org/inews/article.php?story=20060801170800738>.

Weizman, Eyal. interview with Shimon Naveh, *Frieze*, No. 99 (May 2006) <http://www.frieze.com/issue/article/the_art_of_war/>.

Weizman, Ezer. *On Eagle's Wings* (London: Weidenfeld & Nicolson, 1976).

Weller, Jac. "Sir Basil Liddell Hart's Disciples in Israel," *Military Review*, Vol. 54, No. 2 (January 1974), pp. 13–23.

White, Jeffrey B. "Some Thoughts on Irregular Warfare," <http://www.cia.gov/csi/studies/96unclass/iregular.htm>.

Wiener, Erez. "From Embarrassment to Awakening," *Maarachot*, 409–410 (December 2006), pp. 4–19.

William R. Louis and Robert W. Stookey (eds.), *The End of the Palestine Mandate* (London: Tauris, 1986).

Windrow, Martin. *The Algerian War 1954–62* (London: Osprey, 1997).

Wohlstetter, Albert J. "Economic and Strategic Considerations in Air Base Location: Wohlstetter, A Preliminary Review," <http://www.rand.org/about/history/wohlstetter/D1114/D1114.html>.

Wohlstetter, Albert J., Fred S. Hoffman, R.J. Lutz, and Henry S. Rowen. "Selection and Use of Strategic Air Bases," <http://www.rand.org/pubs/reports/2006/R266.pdf>.

Yaalon, Moshe. *A Long- Short Way* (Tel Aviv: Yediot Aharonot, 2008) Hebrew].

Yadlin, Amos and Asa Kasher. "The Ethics of Fighting Terror," *Journal of National Defense Studies* 2–3 (2003), pp. 5–12.

Yair, Yoram. *With Me From Lebanon* (Tel-Aviv: Maarachot, 1990) [Hebrew].

Yaniv, Avner. "The Study of Israel's National Security," in Ian Lustick (ed.), *Books on Israel* (Albany: State University of New York Press, 1988), pp. 63–82.

Yehoshua, Yossi. "Brain Damage to the IDF," *Yediot Aharonot Weekend Supplement*, 12 February 2014.

———. "Declining Values," *Yediot Aharonot Weekend Supplement*, 13 July 2007.

Yogev, Amnon. "Israel's Security in the 1990s and Beyond," *Alpayim* 1 (June 1989), pp. 166–85.

Zeevi, Aharon. "Aviv Ne'urim: The Vision and its Implementation," *Maarachot* 358 (April 1998), pp. 3–6.

Index

"Auschwitz Borders" 115, 134
absorptive imitation 49, 50–51
Abu Nidal 88
Abulafia, Amir 72
active defense 62, 67, 84, 99, 132, 135
Adam, Udi 102
Adamsky, Dima ix, 7, 119
aerial
  bombing 31
  dimension 9, 28, 40, 138
  strategy 45
  vehicles 91
  warfare 12, 27
Agranat Commission 44
airborne forces 45, 94, 106
Alexander, Christopher 54
Alexander, Czar 26
Allon, Yigal 58, 109, 115–117, 126, 134
Almog, Oz 116
AMAN 91, 102, 106
Amnesty International 110
anti-tank guided missile (ATGM) 92
Arab Revolt 39, 70, 71, 117, 133, 134
Aristotle 39, 141, 29
Army Combined Arms Center 53
army
  Austrian 58, 129
  Prussian 7, 8, 17, 34, 121, 126, 145
  Russian 32, 39, 58, 129
  US 7, 22, 48, 58
Ashkenazi, Gabi 57, 106, 110, 137
Assa, Haim 95
attrition 35, 43, 63, 64, 66, 93, 124, 125, 145
  conception 63
  rates 41, 139
  situation xi, 63, 89, 111
  strategy x, 63, 66
  terror-based 66
  war of 16, 23, 63, 97, 125
  warfare 6
Auftragstaktik 50, 121
Augustine Laws 18
Avigur, Shaul 58
Avisar, Eytan 129
Aviv Neurim 53
Awali River 70

Barak program 44, 45, 127
Barak, Ehud 45, 106, 124, 131, 132
Bar-Lev, Haim 122
Barnavi, Elie 43, 44
basic security 63, 83, 86, 86n, 88, 123
Bateson, Gregory 54
Battle
  AirLand 14, 53, 103, 161
  of Bint Jbeil 80
  deep 9, 53
  of Nablus 55
  of Tul Karm 80
battlefield
  decision x, 31, 80, 81
  See also battlefield success; decisive victory
  empty 23, 41, 140
  saturated 23, 41, 140
  success 41, 57, 68, 96–98, 100, 118
battlespace 51, 96
Beaufre, André 4
Beer, Israel 79, 117, 118, 129
Beer, Yishai 91
Begin, Menahem 114, 134
Beirut-Damascus Highway 122
Bekah Valley 97
Bellamy, Christopher 23, 140
Ben-Artzi, Efraim 59
Ben-Eliyahu, Eytan 89, 125
Ben-Gurion, David 43, 46, 48, 83, 108, 114, 128, 129, 133
Ben-Israel, Isaac 62, 79, 99, 106
Ben-Moshe, Tuvia 106, 119
Ben-Reuven, Eyal 102
Berger, Thomas 6
Bernhardi, Friedrich 16, 93, 118n
Biddle, Stephen 20
Bitzuism 115, 116, 144
Blitzkrieg xi, xii, 6, 23, 31, 35, 63, 92, 96, 97, 111, 114, 135, 143
Bond, Brian 116, 117
Boyd, John 107
Bülow, Friedrich 27, 30

Campaign
  Gallipoli 58
  Just Reward 101
  Syrian 39
Campaigns between the wars 68
casualties xiii, 18, 20, 21, 41, 51, 73, 83, 84, 98, 99, 112, 118, 143
See also casualty aversion; casualty rates
casualty aversion xiii, 18, 63, 83, 132
casualty rates 9, 17–19, 41, 139
casus belli 86
See also casi belli
casi belli 86
center of gravity 21, 51, 52, 95, 122
Chen, Amatzia 117
civil disobedience 70, 71, 88, 124
civil liberties xii, 30, 42, 84, 90, 141
Clausewitz, Carl von 3, 9, 12, 13, 16, 26,27, 29, 30, 32, 35, 38–40, 45, 95, 118, 119, 138, 140, 141, 146, 161
Code of Ethics 84, 90, 133, 136
Cohen, Eliot ix, 19, 44, 59
Cohen, Eliyahu 58
Cohen, Stuart 63
collateral damage xii, 19, 73, 83, 84, 99, 112, 132, 143
Colleges
  Air Force Academy 53
  Army War College 53
  College of International Security Affairs (CISA) 127
  Combined Arms Center 53
  Command and Staff College xvii, 77
  Joint Services Command and Staff College 146
  King's College 143
  Narine Corps School of Advanced Warfighting 127
  National Defense College xvii, 64, 72, 77
  National War College 127
  School for Advanced Air Power and Space Studies 146
  School of Advanced Military Studies 127
  Tactical Command College 61, 77, 85
  Joint Services Command and Staff College (JSCSC) 146
command and control xii, 12, 16, 19, 34, 45, 77, 104, 112, 121, 144
command system 9, 16, 50, 105, 121, 122
Computer Service Directorate 131
concentration of fire 95, 112
concentration of forces xii, 95, 112
conflict
  Angola 20n
  Azerbaijan 20, 21n
  El Salvador 20n
  Haiti 21, 22
  Rwanda 20
  Zaire 20
counterinsurgency (COIN) 19, 22, 52, 67, 84, 90, 95, 109, 117
counter-terror 103, 135
Cromwell, Oliver 117
cult of technology xi, xii, 6, 10, 11, 40, 43, 79, 107, 112, 137, 138, 143, 144
cult of the offensive xii, 40 114, 115, 118
culture
  military 3, 5, 7, 40, 119, 138
  strategic ix, 5, 7, 83, 119, 128
current security 63, 64, 75, 86, 86n, 92, 123
Czech-Egyptian arms deal 108

Dado Center 57, 66
Dayan, Moshe 109, 118, 119, 121
Debel (village) 92
decisive victory xi, 63, 66,73, 80, 81, 96, 97, 99–101, 105, 112, 114, 124, 143
  See also battlefield decision; battlefield success
deep penetration operations 33
defensible borders 115
Deleuze, Gilles 74
density theory 21, 22
Desch, Michael 5
deterrent operations 68
discriminate use of force xii, 30, 42, 84, 112, 141, 144
Dixon, Norman 32, 68
Doctrine xiv, 1,2,10, 24, 31, 36, 37, 42, 44, 45, 46-47, 50, 52, 53, 58, 59, 63, 64, 73, 117, 127, 137, 141–143, 145, 154, 155
  See also military doctrine
Doctrine and Education Branch (TOHAD) 44
doers, vs. thinkers 2, 33, 42

dominance in battlefield knowledge (DBK) 93
Dori, Yaakov 58
Douhet, Giulio 4, 14, 15, 28, 28n, 31, 33, 42, 141, 142
Druze state 135
Dubno, Yitzhak 58
Duffield, John 5, 8

Eban, Abba 115
effect-based-operations (EBO) 51
Eisenhower, Dwight D. 33
Eisenkot, Gadi 68, 81
Eitam, Ephraim 132
Elazar, David (Dado) 57, 122
Engels, Friedrich 31, 38, 39, 141
Enlightenment 28–30, 38, 140, 141
entrenched traditionalism 8, 24, 33, 36, 40, 42, 124, 138, 141, 142, 145
Eshel, Amir 81, 103
Eshet, Fritz 58
ethnic conflict 24, 25, 42, 140
ethnocentrism 32, 32n
ethno-religious coalition 130
excessive use of force 19, 99, 110
Eyval Gil'adi 80

Fajr (rockets) 98
Farrell, Theo 8
Firepower 13, 14, 17, 41, 51, 62, 68, 81–85, 92–101, 112, 143
Firestones-9 88
first strike xii, 31, 92, 96, 107–109, 112
Foch, Ferdinand 4
focused logistics 112, 143
Force 17 56
force multipliers xii, 22, 31, 43, 81, 85, 92, 112, 143
force-to-space ratio 21, 22
Forester, John 54
Foster, Gregory 48, 116
Frederick the Great 16, 26, 29–31, 45, 140
French Revolution 12, 27
Fuller, John F.C. 4, 7, 10, 13, 27, 38, 93, 117

Galilli, Elazar (Lasia) 58
Gallois, Pierre Marie 4
Gandhi, Mahatma 70
Gaza 76, 77, 89, 98, 126
Geertz, Clifford 54
General Security Services (GSS) 91, 106
Glenn, Russell W. 74
Globerman, Yehoshua 58, 129
Gneisenau, August von 16
Gorbachev, Mikhail 127, 146
Gordon, Shmuel 103
Grad (rockets) 98
grand-strategization of tactics 16, 77, 85
Gray, Colin 6, 7
great debates 43, 57, 61, 85, 142, 145
Green, Mordechai (Monti) 58
ground forces digitalization program 104
Guattari, Félix 74
guerrilla 22, 39, 99
Guibert, Jacques 4, 15, 26, 29, 30
Gur, Mordechai (Mota) 121

HaCohen, Gershon 119
Haganah 45, 57, 58, 114, 133
Halutz, Dan 71, 80, 89, 101, 102, 104, 125, 133, 134
Hamas 98, 126
Hammes, Thomas 25, 140
Handel, Michael 11, 26
Hannibal 120
Harris, Fred 58
Hegel, Georg 39, 141
Henry, Duke of Rohan 38, 140
Hezbollah 80, 81, 89, 91, 92, 94, 95, 98, 101, 103
Hirsch, Gal 56
Hisdai, Yaakov 44
Hoffman, Frank 11, 25, 139, 140
Horowitz, Dan x, 119, 127
Howard, Michael 9, 14, 40, 138, 139
Hubris xiii, 115, 122, 137, 144
Human Rights Watch 110
Huntington, Samuel 8

Improvisation ix, xiii, xvi, 7, 17, 45, 114, 115, 118–121, 137, 138, 144
Industrial Revolution 12, 27, 118
Information
dominance xii, 11, 112, 143
glut 9, 11

Institute for National Security Studies (INSS)105
institutional military xiii, 34, 128
intellectual
  pretense x, 43, 53, 85, 102
  soldiers xiii, xv, 4, 36, 37, 42, 127, 145
intellectualism
  anti- ix, xii, 12, 43, 44, 54, 119
  false x, 43, 53, 8, 142
  institutional xiii, 126, 137, 144, 145
intelligence
  field 45
  Human (HUMINT) 91
  Signals (SIGINT) 91
International Law Department (ILD) 136
International Law Unit 110
Intifada xi, xv, 63, 76, 84, 89, 130, 136, 143
  First 70, 71, 85, 88, 90, 111, 124, 142
  Second 887, 91, 135, 136
Intuition x, xiii, 2–4, 36, 50, 93, 94, 115–118, 135, 137, 145
Islamic Jihad 126
Israel Air Force (IAF) 45, 49, 51, 52, 68, 71, 72, 80, 89, 91, 97, 101–103, 120, 125, 131, 133
Ivry, David 72n, 102

Janowitz, Morris 8, 37
Johnston, Alastair 6, 6n 8
jointness 52, 67, 117, 146
Joint Chiefs of Staff 93n
Jomini, Antoine-Henri 3, 12, 27, 30, 32, 35, 38, 38n, 39, 93, 119, 140–142
judicialization 84, 114, 136, 137
jus ad bellum 107, 108n
jus in bello 107, 109

Kaldor, Mary 25
Kant, Immanuel 39, 141
Kaplinsky, Moshe 89, 125
Katyusha 72n, 102
Katznelson, Berl 133
Kennedy School of Government 53
Kibbutz 58, 128, 130
Knesset 127, 132, 145
Knesset Foreign Affairs and Defense Committee 77, 79, 102
Knott, Steven 2, 33, 34, 37
Kokhavi, Aviv 55
Krauthammer, Charles 19

Laskov, Haim 59, 129
Laurence of Arabia 39, 117, 142
legal advisers 84, 110, 111, 137
  See also military lawyers
legal considerations 82, 110, 132, 136, 143
Legro, Jeffrey 53
Lendon, J.E. 9
Lenin, Vladimir 31
lesson learning 34, 42, 64, 88, 90, 120, 141
levels-of-war x, xi, xiv, xvi, 4, 15, 16, 41, 66, 76, 78, 89
leverages and effects xi, 89, 112, 143
Lev, Rephael 129
Levin, Amiram 68, 68n
Levy, Yagil 63, 130
Liddell Hart, Basil H. x, 4, 13–15, 27, 31, 36, 38, 45, 93–95, 116, 117, 139–141
Lilienstein, August 4
limited confrontation 64, 66
Lind, William 25, 140
Lipsius, Justus 28, 38
Lissak, Moshe 127
Livni, Tzipi 81
Lloyd, Henry 38, 140, 141
logic of the few 95
logistical dimension xiv, 12–14, 27, 35, 43, 70, 72, 80, 82, 85, 91, 96, 112, 116, 143, 158, 159
low-intensity conflict (LIC) ix, x, xi, xii, xiii, xiv, 16, 18–25, 28, 41–43, 59, 64, 66–68, 73, 75–77, 80, 81, 83, 85, 86, 88–90, 97–100, 107, 111, 112, 115, 116, 118, 122–125, 132, 135, 137, 139–144, 156, 161
Luftwaffe 50
Luttwak, Edward x, 2, 49n, 83, 119, 132, 140

Machiavelli, Niccolò 28, 38
Maginot Line 10
Mahan, Alfred Thayer 4, 13, 27, 39, 45, 141
Maklef, Mordechai 108
Mao tse-tung 4, 6n, 31, 38, 141
Marcus, David 58, 59
Marx, Karl 27, 31, 38, 141
Metz, Steven 23, 139
Middle Ages 11, 28
Military Advocate General 110, 135

military buildup xii, 10, 13, 14, 24, 43, 61, 79, 81, 85, 123
military doctrine ix, 49, 53, 79, 99, 100, 103, 110
  AirLand battle 53
  American xi, 43, 50, 59, 66
  Arab 70
  Biblical 117
  current security 64
  Dahiyya 68
  deep battle 53, 116
  emulated 43, 50, 59, 72, 73
  false 3, 46
  IAF 120
  LIC 124
  mission command ix, 7
  operational 59, 73, 97
  Powell 18
  revolutionary 31
  RMA-inspired x, 73, 85, 112, 142
  Soviet 70
  strategic ix, 96
  tactical 59
military genius 3, 16, 29, 30, 33, 141
military intellectuals xiii, xiv, xv, 34, 36, 37, 42, 56, 79, 128, 129
military lawyers xi, xiii, 43, 84, 110, 112, 137
  See also legal advisers
military mind 33, 42, 141
military thought
  American xi, xvi, 51, 126, 149
  British 58, 129
  German 16, 34, 35, 58, 121, 126, 129, 145
  Prussian 7, 8, 17, 31, 34, 121,126, 145
  Soviet 7, 14, 34, 53, 54, 70, 126, 129, 145
Military Review xi, xv, xvi, 48, 143, 149, 151, 154, 156, 158, 159, 161
Military Technological Revolution (MTR) 10, 53
Miller, Benjamin 42, 140
Milstein, Uri 44, 117
mission command ix, 7, 16, 17, 96, 121, 122
  See also mission command doctrine
Mitchell, William (Billy) 4, 14, 15, 28, 33, 142
Mofaz, Shaul 51n, 54, 88, 59, 115, 124, 126
Moltke, Helmuth von 4, 16, 17, 31, 38, 39, 93, 118, 121, 141
Montecuccoli, Raimondo 29, 30
Montesquier, Charles-Louis de Secondat 39, 141
mowing the grass 68

Nahushtan, Ido 71
narco-terrorism 23, 105, 140
National Security Studies Center 64
National Defense University (NDU) 53, 127
Naveh, Shimon 54, 55
neo-classicism 28
New Wars 25, 46
Ney, Michel 39
night squads 58
nuclear, revolution 10

observation-orientation-decision-action loop 103
Olmert, Ehud 89, 94, 102, 125, 126
Operation
  Cast Lead 98, 110
  Defensive Shield 56, 88, 89, 126
  Exporter 70
  Mey Marom 102
  Pillars of Defense 98, 110
  Protective Edge 110
  Qibya 109
operational art x, 54, 73, 81, 82, 85, 92, 94, 96, 143
Operational Theory Research Institute (OTRI) xiii, 54, 56, 57, 73, 74, 76
Operational Code 36
operational dimension 9, 10, 14, 79, 80, 158
operations other than war (OOTW) 64
Ott, Steven 8

Palestinian Authority (PA) 56
Palmach 58, 118, 127, 133
Patton, George 2, 33
Pedatzur, Reuven 62
Peled, Benny 131
Peled, Yossi 101
Peres, Shimon 86
Peretz, Amir 101
Picq, Ardant du 39
planning xiv, 2, 56, 127, 136, 145, 154, 155
  strategic 23, 37, 46–47, 50, 80, 81, 96, 140
  operational 110, 111, 116, 137
Plato 29, 39, 141

Poirier, Lucien 4
policing missions x, 85, 89–91
popular republics 28, 29
Posen, Barry 8
Postman, Neil 10
post-modern thought 4, 36, 55, 143
practical soldiers x, xv, 36, 37, 42, 117, 145
precision weapons 9, 18, 19, 51, 62, 73, 84, 97, 99, 104, 112, 132, 143
principles of war 30, 38, 120, 140
proportionality xii, 30, 42, 68, 84, 90, 105, 107n, 112, 141, 144
purity of arms 90, 109, 133, 136

Qassam (rockets) 98, 126
Quinlivan, James 21

Rabin, Yitzhak 88, 94, 116, 117, 121, 124
Rand Corporation 2, 21, 74
Ratner, Yohanan 58, 129
reflective practitioners 37
Research and Development Directorate 99, 106
Revolution in Military Affairs (RMA) 41, 43, 51, 53, 62, 73, 85, 89, 95, 99, 112, 119, 122, 139, 142, 143
romanticism 28–30, 140, 141
Ron-Tal, Yiftah 91
Rosen, Stephen 7
Roszak, Theodore 11
Rousseau, Jean Jacque 39, 141
Rubin, Doron 44

Sagi, Uri 68
Saxe, Morris de 26, 29
Scholars, vs. warriors 34, 37
School for Advanced International Studies (SAIS) 53
Schueftan, Dan 62
security conception x, 88, 106
sensor-to-shooter loop 51
Seversky, Alexander 14, 15, 28
Shamir, Eitan ix, 7
Shamir, Shlomo 58
Shamir, Yitzhak 88
Sharet, Moshe 108
Scharnhorst, Gerhard von 16, 34, 146
Sharon, Ariel 77, 86, 88, 94, 115, 124
Shkedy, Eliezer 72n, 89
Shomron, Dan 99
Silva, Marquise de 38, 140
small and smart military 21
societal dimension 12
soldier-statesman 77
space 55, 56
  battlespace 51, 96
  cyber 40, 105, 106, 138
  fifth 105
  open 120
  outer 40, 105, 138
  protected 90
  smooth and striated 55
  warfare 9, 12
Special Forces 51, 103
spiritual father 36, 38, 42, 54
squaring the triangle 9, 13, 138
stability operations 21
stand-off-weapons 84, 101, 104
State Comptroller 50
state-to-nation balance 140
Strategic Air Base Study 2
strategic studies 41, 139
strategization of tactics x, 4
strategy, dimensions of 9, 14, 40, 79, 80, 138, 158, 159
Suez Canal 94, 120
Sun Tzu 38, 45, 95
surface-to-air missiles 98, 120
surface-to-sea missiles 98
surface-to-surface missiles 103
swarming 52

tacticization of strategy x, 4, 19
Tal, Israel 62n, 68, 93, 115, 134
Tamir, Moshe (Chico) 80
targeted killing 68, 84, 91, 110, 135, 137
technowar 11
Tolstoy, Leo 36
troop density paradox 21, 22
Tuchachevsky, Mikhail 33, 142
Tziklon (Journal) 45, 69, 70

Unit 8200 106, 131
unmanned aerial vehicles (UAV) 91
US Marine Corps 2, 34, 127, 146, 151n
Vagman, Yehuda 66
Van Buren, Dany 100

Van Creveld, Martin x, 20, 25, 42–44, 72, 90, 91, 121, 122, 140
Vauban, Sebastian de 26, 30
victory from the air 71, 100, 102
victory image xi, 42, 80, 140, 143
victory show 80, 112
Vilnai, Matan 104
Voroshilov, Kliment 33

Walzer, Michael 109
war
  Afghan 19, 19n, 20n, 21, 21n, 22, 24, 70
  Algerian 39, 66
  American Civil 13, 35
  art of 3, 12, 38, 146
  asymmetrical xi, xii, 11, 21–24, 63, 63n, 64, 81, 88–90, 99, 100, 111, 123, 132, 143
  Boer 35
  broadening of 11–15, 17, 22, 41, 139
  complexity of 16, 22–24, 40, 41, 48, 54, 139
  of consciousness 88
  of conviction 112, 63, 88
  Crimean 39
  First Lebanon xi, xv, 44, 69, 94, 102, 112, 135
  First World 9, 12, 13, 16, 21, 23, 27, 28, 34, 35, 40, 58, 122, 138
  Gaza *See Operation Protective Edge*
  Gray Area 23, 42, 139
  Hybrid 23, 25, 41, 42, 46, 66, 90, 139, 140
  Independence xv, 43, 94, 114, 116, 117
  Indochina 24, 66
  Iraq 17, 19, 19n, 21, 22, 109
  just xii, 8, 42, 84, 107, 107n, 109, 134, 135, 141
  Korean 35
  Kosovo 18, 19n, 21, 71
  morality and xi-xiii, 10, 25, 26, 30, 63, 110, 114, 135
  narrowing of 14, 17, 20, 22, 41, 139
  non-trinitarian 25, 42, 46, 140
  nuclear 2, 10, 12, 28, 35, 45, 68, 75, 105, 106, 109, 144
  October xv, 44–45, 48, 57, 61–62, 64, 67, 93–94, 97, 102, 120, 129
  post-heroic xiii, 1, 42, 63, 68, 82, 83, 85, 92, 114, 132, 133, 139, 140
  preemptive 107, 108, 109
  preventive 107, 108, 109
  regular 23, 25, 58, 76, 89, 123, 124
  revolutionary 24
  Russo-Japanese 35
  Russo-Turkish 35
  Second Lebanon xii, 22, 44, 57, 66, 68, 71–74, 79, 81, 82, 84, 85, 88–91, 94, 96, 99–104, 111, 112, 122, 125, 142
  Second World xii, 2, 9, 12, 13, 16, 17, 20, 21, 23, 28, 30, 34, 35, 41, 45, 58, 70, 117, 118, 129, 139, 140
  Seven Years' 16, 122
  Sinai xv, 108, 114, 118, 119
  Six-Day xv, 96–97, 109, 114, 119, 122
  sub-conventional xiv, 4, 24, 65, 75, 76, 87, 157, 158
  unconventional xiv, 4, 65, 75, 76, 87, 156–158
  way of xiii, 6, 82, 83, 132, 133
Warfare
  Air 1, 9, 10
  anti-tank 120
  complex irregular warfare 25, 66
  cyber 9, 10, 12, 105–107, 112, 131, 144
  Fourth Generation 25, 41, 42, 46, 66
  diffused 51, 95, 96
  mountainous terrain 1, 45, 69, 70
  naval 1, 32, 39, 45, 141
  network-centric 51, 103, 104, 144
wars
  laws of 110, 136
  Napoleonic 12, 26, 27, 34, 35, 39, 140
  New 25
Weapons
  anti-tank 45
  biological 75
  chemical 75
  nuclear *See nuclear war*
Wingate, Orde 58, 117
Winograd Commission 44, 56, 74, 84, 101, 111, 119

Yaalon, Moshe 56, 81, 99, 102, 124, 125
Yaari, Yedidya 95
Yadin, Yigael 108, 129
Yadlin, Amos 71, 102, 106
Yair, Yoram 70, 92, 94
Yogev, Amnon 97

Zeitgeist 25, 40, 42, 138, 140
Zion Mule Corps 58

Printed in the United States
By Bookmasters